W9-BNT-493

THE ESSENTIAL BOOK OF WEATHER LORE

THE ESSENTIAL BOOK OF WEATHER LORE

Time-Tested Weather Wisdom and Why the Weatherman Isn't Always Right

LESLIE ALAN HORVITZ

The Reader's Digest Association, Inc.
Pleasantville, NY/Montreal

A READER'S DIGEST BOOK

A Marshall Edition
Conceived, edited and designed by Marshall Editions
The Old Brewery
6 Blundell Street
London N7 9BH
England

This edition published by The Reader's Digest Association, Inc., by arrangement with Marshall Editions.

FOR MARSHALL EDITIONS
Publisher: Richard Green
Commissioning Editor: Claudia Martin
Art Director: Ivo Marloh
Editor: Deborah Hercun
Design: 3rd-I
Indexer: Lynn Bresler
Production: Anna Pauletti

Cover originated in Hong Kong by Modern Age

FOR READER'S DIGEST
U.S. Project Editors: Kimberly Casey, Mary Connell
Copy Editor: Barbara McIntosh Webb
Canadian Project Editor: Pamela Johnson
Canadian Consulting Editor: Jim Hynes
Associate Art Director: George McKeon
Executive Editor, Trade Publishing: Dolores York
Associate Publisher: Rosanne McManus
President and Publisher, Trade Publishing: Harold Clarke

Library of Congress Cataloging in Publication Data:
Horvitz, Leslie Alan.
The essential book of weather lore: time-tested weather wisdom and
why the weatherman isn't always right / by Leslie Alan Horvitz.
p. cm.
Includes index.
ISBN 13: 978-0-7621-0857-2
1. Weather forecasting—Popular works. 2. Weather—Folklore. I. Title.
QC995.4.H665 2007
551.63'1--dc22

2007013354

We are committed to both the quality of our products and the service we provide to our customers.
We value your comments, so please feel free to contact us.

The Reader's Digest Association, Inc.
Adult Trade Publishing
Reader's Digest Road
Pleasantville, NY 10570-7000

For more Reader's Digest products and information, visit our website:
www.rd.com (in the United States)
www.readersdigest.ca (in Canada)

Printed and bound in China by Midas Printing International Ltd

1 3 5 7 9 10 8 6 4 2

CONTENTS

Introduction

Weather folklore is often dismissed as nothing more than a grab bag of sayings, old wives' tales, legends, and superstitions. In other words, folklore is considered the opposite of science. But folklore and science have more in common than you might imagine. What we call the scientific method is based on observation and evidence—and so is a great deal of weather folklore.

Weather lore arose in response to a problem that confronts us today: how to predict the weather with any degree of accuracy. Ancient peoples were anxious to learn what weather to expect, just as we are today, because their very survival depended on it. Farmers looked to nature for clues as to what the growing season would bring; sailors scanned the skies and took note of wind direction before embarking on a long voyage.

In the absence of instruments such as thermometers and barometers, people were forced to rely on their senses. Many of the observations they made involved associations: Red skies at night followed by gray skies the next morning indicated a fair day ahead. Animals, insects, and plants were pressed into service as forecasters because so many of them demonstrated certain types of behavior before abrupt changes in weather. For instance, some flowering plants, such as marigolds or crocuses, will close before a rainstorm because of increasing humidity. Once these observations proved consistently reliable over time and in different parts of the world, they were incorporated into a body of proverbs, sayings, and verse that collectively make up weather folklore. Many weather lore sayings are indeed just that—folklore indebted to the oral tradition, just like the legends and tales of old. Inevitably, these sayings underwent many changes as they were passed from one generation to another and traveled from one region to another, generating multiple variations and distortions along the way. Nonetheless, many sayings from widely scattered parts of the world are remarkably similar.

Weather lore, however, has only limited predictive value. The forecasts of the celebrity groundhog Punxsutawney Phil, for example, are accurate about 50 percent of the time, producing results no better than flipping a coin. Moreover, these associations are just that: They do not account for why an event occurred; "If A happens, so will B" may mean that A causes B, but it also may mean that

A and B are caused by C, or that A occurs before B only by coincidence. How weather happens depends on an understanding of complex phenomena, such as the makeup of Earth's atmosphere and the behavior of fronts, jet streams, and air masses, knowledge that could only be acquired with the development of technology—for example, weather satellites, Doppler radar, and fast computers.

In addition, weather lore is constrained by geography. Until the sixteenth and seventeenth centuries, people had no tools or methods at their disposal that would allow them to find out what weather conditions were like in a distant location. But accurate weather forecasting is impossible unless such information is made available in a timely fashion. You may be basking under blue skies today, but a storm upwind may well end up soaking you tomorrow.

Thunderstorms over Florida
Florida is among the places most frequently struck by lightning on Earth. It is surrounded by water, to the east by the Atlantic Ocean and to the west by the Gulf of Mexico, creating clashing sea winds off the water that cause severe thunderstorms down the spine of the peninsula.

1

SEASONS AND CLIMATE

Trying to get a jump on the year is an enduring preoccupation. In using natural phenomena to forecast weather for a season or to predict how the entire year will turn out, folklorists have proven far more intrepid than meteorologists, who usually confine their forecasts to a period of four or five days in advance. Some of the folklore methods are simply superstition, with no basis in fact; others, relying on decades or centuries of observation, may have some limited validity. One thing is certain, though: For those who do believe that that they can predict the weather far into the future, past failure is no deterrent.

THE YEAR AHEAD

Traditionally, the waning days of the old year or the first days of the new had special significance because they were said to foretell how the upcoming year would unfold. The dilemma in using this method is obvious. How is it possible to rely on weather conditions over a period of 12 or so days to predict what the weather will be like in the warmer months?

Dirty days hath September
April, June, and November
From January up to May
The rain it raineth every day
All the rest have thirty-one
Without a blessed gleam of sun
And if any of them had two-and-thirty
They'd be just as wet and twice as dirty.

As a sweeping generalization, this whimsical bit of poetry leaves something to be desired, but for certain regions of the world it holds a kernel of truth. It applies especially to winters on the west coasts of both Europe and North America. Soggy winters tend to be the rule in London and Dublin as well as in Vancouver and Seattle (which endured the wettest November in its history in 2006).

CALENDAR

Calendars have been used for thousands of years to mark feast days, harvests, and the beginning and end of the year. The oldest calendars are lunar and calculate the period from one new moon to the next—a so-called lunation. The ancient Egyptians used a lunar calendar with 12 months of 30 days each. It is estimated that there are about 40 calendars based on different systems in use today. The Christian or Gregorian calendar is based on the orbit of Earth around the sun. The Islamic calendar is based on the lunar cycle with no reference to the sun; the Jewish calendar combines both systems.

A clear and bright sun on Christmas Day fortelleth a peaceable year and plenty; but if the wind grow stormy before sunset, it betokeneth sickness in the spring and autumn quarters.

In Europe, what the weather was like on each of the 12 days of Christmas was said to predict weather conditions likely to occur in each month of the new year. (Some cultures began the countdown with the winter solstice, a few days before.) So the weather on Christmas Day would provide you with a glimpse of the weather in January, while the eighth day of Christmas would prepare you for August. Similarly, a wet Christmas season would serve as a warning that the whole year was going to be wetter than usual. Needless to say, this belief has no basis in scientific fact.

The nearer the New Moon to Christmas Day, the harder the winter.

There is no correspondence between the phase of the moon and the weather that can be used with any reliability to make weather predictions a few days or a week ahead of time, let alone an entire season. Weather varies considerably from one place to another and from one time of day to another, while the moon is in the same phase for everyone wherever they happen to live. Moreover, the phase of the moon changes only slowly. This is not to say that the moon has absolutely no effect on the weather: It does have an important effect on tides, so the lunar phases have an indirect affect on local weather, especially in areas close to the ocean. Tides affect the mixing of the seas, churning up the water. In tropical seas, a great deal of mixing will cool the water. In Arctic or Antarctic waters, which are often covered with ice, there will be little mixing because ice acts as an insulator.

When there are lots of berries on the dogberry tree, it means it's going to be a bad winter!

The dogwood tree, cornelian cherry, native to southern Europe and the Orient, blooms in February and March, bursting into masses of yellow blossoms. In late summers, it produces scarlet berries that remain on the branches until September, when the birds eat them all. If the berries have remained on the tree late in the season, as this saying indicates, it would imply that birds had migrated very early to escape the harsh winter ahead.

Flowering Dogwood

The Dogwood family is made up of 30 to 50 species of trees and shrubs, some of which produce white flowers. "Dogwood" is a corruption of "dagwood," which refers to the stems of the hard wood used to make "dags" (daggers or skewers). Dogberry is a lesser-used term. According to legend, the dogwood was used to construct the cross on which Christ was crucified; the reddish center of the flowers symbolizes Christ's blood.

ETRUSCAN SOOTHSAYERS
The Etruscans, whose civilization flourished in northern Italy around 9000 B.C., practiced a form of divination known as haruspicy. To make their predictions, the soothsayers—called haruspices (diviners)—would examine the entrails of animals. They used the weather in their prophesying: It is thought that they could foretell disaster or good fortune by measuring changes in wind direction, claps of thunder, or bolts of lightning.

When leaves fall late, winter will be severe.

The reasons for leaf fall are still not entirely understood. Some authorities believe that it is a form of plant excretion; others theorize that the shedding protects against diseases and pests. Dead leaves that linger on branches until late in the season may help protect trees against freezing conditions. But there's little reason to think that a late leaf fall can predict whether winter will be mild or severe.

If at Christmas ice hangs on the willow, clover may be cut at Easter.

Although there is no proof that a frigid Christmas will bring a warm Easter, it is understandable why Christian tradition would associate the two holy days. The aromatic willow tree, found in temperate regions and near waterways, is said to have both medicinal and magical uses. Ancient tradition associated willows with bees because they have been used as honey trees. More to the point of this saying, willows are associated with mistletoe, a plant that brings to mind Christmas rituals. The willow was considered one of the Seven Sacred Irish Trees and is also a revered Druid Tree. Seasonally, the willow keeps its foliage well into autumn, long after most trees have shed their leaves. They also blossom earlier than other trees, making them a sign of spring along with the robin. That they bloom so early in spring might also account for why willows symbolized the Resurrection at Easter, while at the same time connecting them to the colder weather of Christmastime.

Onion skins very thin, mild winter coming in; onion skins thick and tough, coming winter cold and rough.

If anything, the thinness or thickness of onion skins is more of an indication of conditions in which the onions were grown than a predictor of future weather patterns. Nonetheless, onions enjoy a rich history. Onions are one of the few vegetables that can keep over winter. According to some traditions, onions can also be used to predict the amount of precipitation to expect

Onions
Many folk remedies have been attributed to the onion. An onion under a pillow is supposed to cure insomnia, and chewing an onion is said to prevent colds and sore throats.

in the coming year: The onion was known as far back as 3500 B.C. More than just a food, it was an ancient symbol of eternity because of the concentric circles that are revealed once it is sliced. Ancient Egyptians worshipped the onion because of its association with eternity, which also supposedly accounts for why Russian Orthodox churches place domes on their places of worship in the shape of the onion.

According to a Turkish legend, when Satan was cast out of heaven, garlic sprouted where he placed his left foot and an onion where he placed his right foot. Onions are also associated with December 1, which in some parts of the world was the day for young girls to perform the ancient art of cromniomancy (divination by onion sprouts). This method is supposed to reveal the name of their future husbands. Another legend prescribes how to use the onion to predict what winter will be like: Between 11:00 P.M. and midnight on either Christmas Eve or New Year's Eve, take a dozen onions, cut off their tops, and scoop out a hole in their centers; then line them up in an east-west orientation. Place an equal amount of salt in each hole. The salt will dissolve to different degrees. The more water in each onion, the wetter the corresponding month is supposed to be. But don't look for the results until the next morning.

WEATHER INSTRUMENTS

The modern science of meteorology was made possible by the invention of instruments to observe and measure the weather. The most common such instrument is the thermometer. Its invention is attributed to the Italian scientist Galileo Galilei (A.D. 1564–1642) in 1593. Thermometers measure temperature on three possible scales: Fahrenheit, Celsius (Centigrade), or Kelvin. The most common type is the mercury bulb thermometer in which mercury moves up and down a calibrated tube as it contracts and expands with changes in temperature. Barometers are used to forecast changes in the weather by measuring atmospheric (barometric) pressure. Falling barometric pressure indicates the approach of a low-pressure area associated with storms, while rising pressure means that a high-pressure system is moving in, bringing warmer, clearer weather. An anemometer measures wind velocity and pressure. The hygrometer measures relative humidity; an early version was invented by Leonardo da Vinci (A.D. 1453–1519) in the late 1400s. The rain gauge is the simplest instrument; it is commonly placed at weather stations and airports to measure amounts of rain and snowfall.

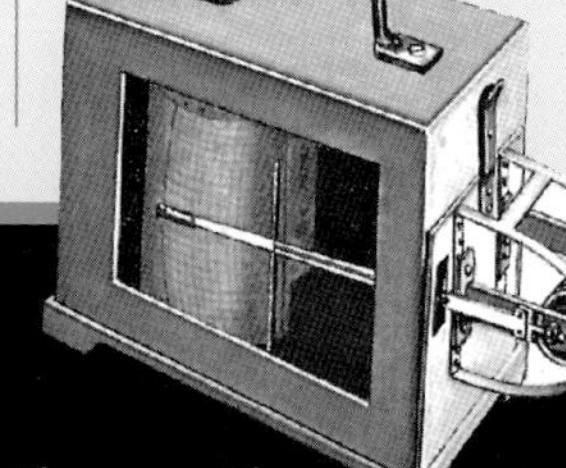

Thermograph: *A thermograph records the temperature on a revolving cylinder made out of a strip of temperature-sensitive metal.*

Dry-bulb thermometer

Barometer

Farming and Astrometerology

According to the adherents of astrometerology, Cancer and Scorpio are considered especially fruitful signs, favorable for germination and for producing abundant harvests.

Frost on the shortest day bodes a bad winter.

Decembers are typically moderate in Wales, where this saying originates. So frost on the shortest day, December 21 (the winter solstice), may well indicate a severe winter. Nearly every year, Wales is buffeted after Christmas by high winds and snow, commonly referred to as the "post-Christmas storm."

When grape leaves turn yellow early in the season, it will be a harsh winter.

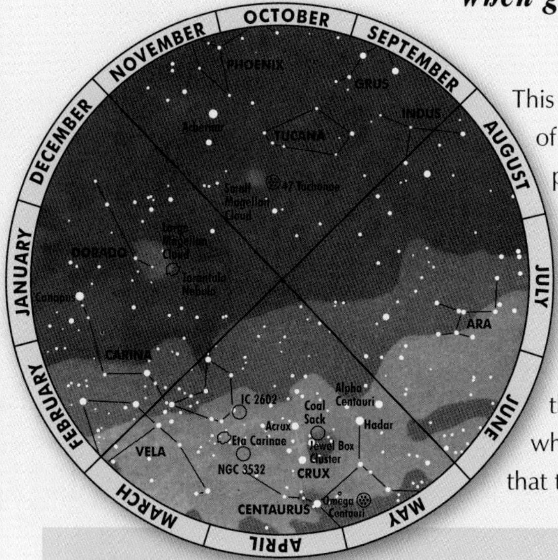

This Canadian saying is true only insofar as it describes the effect of cooler weather on the plant, but it has no predictive value. All parts of the grape cluster (including the cluster stems, as well as the berries) are susceptible to infection by fungi throughout the growing season. Most infections will show up early in the season. Infected berries will turn a light brown color. Once inside the tissues of the berry, the fungi can sometimes become inactive, meaning that the disease does not continue to develop. As a result the infected berries remain asymptomatic until late in the season when the fruit matures. Cool, wet weather increases the possibility that the infection might cause the latent spores to start germinating.

ASTROMETEROLOGY

Astrometerology is the practice of forecasting weather—especially storms, floods, hurricanes, and even earthquakes—based on observations of the sun, the moon, planets, and constellations. Each zodiac sign is supposed to influence planets occupying a particular region of the sky. For example, the moon (referred to by practitioners as Luna) is regarded as a "wet planet," which indicates more precipitation than usual when Luna moves into a water sign, such as Pisces, than when it is located in a fire sign, such as Leo. Adherents of this ancient method contend that they can predict meteorological events or natural disasters years in advance. There is some basis in science for the practice (though it is not taken seriously by most meteorologists). Planets are in fact surrounded by magnetic fields that can extend for millions of miles into space and can affect Earth's atmosphere. Farmers have also used this method. Leafy vegetables that grow above ground, for instance, should be planted during a new or waxing moon in Pisces, Taurus, Cancer, or Scorpio.

Prepare for a hard winter when hornets have triple-insulated nests.

Insects, like animals, are often the subject of weather lore. Insects are believed to anticipate the severity of winter by their behavior. So it would seem to follow that hornets would prepare heavily fortified nests where they could hunker down for months of frigid temperatures. A hornet's nest is created by a single fertilized queen after she emerges from hibernation (in late April in the temperate Northern Hemisphere).

The queen lays a single fertilized egg in each cell of the nest. There they remain until they reach maturity, their development hastened by a diet of spiders. Almost all hornets are females. They then go to work helping the queen build more cells and feed and care for a new generation. As the weather gets colder, the queen dies and her colony disperses. Any surviving hornets that choose to stay behind are killed by the first cold snaps. Occupied nests are seldom seen after early autumn, and what remains of them usually disintegrates in the winds and ice of winter. Hornets never return to an old nest and are unlikely to rebuild in an area that their forebears colonized. So it would make no sense for hornets to waste their energy adding insulation to a nest that will be of no use within a month.

Hornet's Nest

The female hornet builds her nest out of "paper" cells—actually chewed tree bark—that are arranged in horizontal layers named combs. Each cell is vertical and closed at the top. As the colony expands, new cells are built, rather like an extension on a house.

A tough winter is ahead if birds migrate early.

There is some truth to this saying, but it is limited mainly to species of birds that migrate over relatively short distances. Long-distance migrants, which include large broad-winged birds such as vultures, eagles, and buzzards, leave for warmer latitudes in late fall because they are genetically programmed to respond to the changing lengths of days. Since weather conditions have no effect on the length of the day, these birds won't leave any earlier, in spite of colder or windier conditions than normal. But birds that tend to migrate for shorter distances will leave earlier than usual if the weather becomes especially harsh. These birds include mountain and moorland breeders such as wallcreepers, merlins, and skylarks. In very cold weather the chaffinch, which is native to Britain, will migrate to the south of Ireland. In North America, some adult male Ruby-throated Hummingbirds head south as early as July 1, while the juveniles may not leave until November.

EARTHQUAKE WEATHER

Although there is scant evidence that weather can predict earthquakes, the belief that such a link does exist has persisted since antiquity. Geologists generally dismiss the idea that weather has anything to do with earthquakes, pointing out that weather affects the surface of the Earth, while seismic activity takes place deep underground. Moreover, earthquakes occur in all kinds of weather and during all seasons of the year.

William Shakespeare

The great playwright was acutely aware of the effects of weather on his characters. In some plays the weather influences the plot, as in The Tempest. *In others, the weather reflects and elaborates upon the temperament and plight of the characters.*

A windy winter, a rainy spring.

There is some truth to this saying, although its accuracy depends on which region of the world it applies to and from where the wind is blowing. In Scandinavia, for instance, and in Norway in particular, the weather is very much governed by the direction of the wind. Southerlies and easterlies bring sunny weather, while westerlies bring precipitation with mild weather in winter and cool rainy weather in summer, if it has not arrived by spring. Northwesterlies, however, bring the worst weather, with snow or sleet in the winter along the coast.

When oak is out before the ash,
'twill be a summer of wet and splash.
But if the ash before the oak,
'twill be a summer of fire and smoke.

This English rhyme refers to the use of the two beloved tree species as barometers. This saying is often paired with a variation that offers a forecast of rain regardless of which tree blooms first: "If oak's before ash, you're in for a splash. If ash before oak, you're in for a soak." The former may apply only in southeast England, where the bud-break of the oak almost always occurs about two weeks before the bud-break of the ash, regardless of the weather conditions. In other parts of the country, it has been a kind of contest as to which of the two would break its buds first. However, conservationists are concerned that climate change is altering the "rules" of the contest, with oaks almost invariably coming into leaf earlier than ash because they are responding to the changes more rapidly—about 10 days earlier than they were

SHAKESPEAREAN WEATHER

An astute observer of human nature, poet and playwright William Shakespeare (1564–1616) was aware of the effect of seasonal changes on his characters, and weather plays a key part in many of his plays. In *King Lear*, for example, violent storms reflected dramatic events onstage. The opposite holds true, too: The emotional turmoil of the characters could often have an impact on the weather.

In *A Midsummer Night's Dream*, for instance, the estrangement of Oberon and Titania is reflected by the following:

" ... the winds, piping to us in vain,
As in revenge, have suck'd up from the sea
Contagious fogs; which falling in the land
Have every pelting river made so proud
That they have overborne their continents."

two decades ago. This claim is based on findings by the Woodland Trust, the United Kingdom's leading woodland conservation charity, and the Centre for Ecology and Hydrology in Cambridgeshire. This would not be a source of great concern were it not for the fact that climate change may be endangering the survival of the ash and other tree species. The fear is that oaks, together with other early budding trees such as sycamores and horse chestnuts, would enjoy a competitive advantage. Their early leafing would cast shade over the later-budding trees. As a result, fewer ash trees would grow or else would disappear entirely from certain regions. The upshot is that in the future, ash may never bud before oaks.

A rainy spring, a severe summer.

The idea that weather conditions prevailing in one season (or one month or one week) could be related to conditions in another falls under the rubric of phenology. (Phenology is not to be confused with phrenology, which is based on the belief that the contours of the head provide clues to an individual's personality.) Phenology is defined as the study of recurring natural phenomena. The word is derived from the Greek *phainomai*, which means "appears" or "comes into view."

Natural cycles are a major focus of study, especially the dates that mark the first appearance of natural events—the first frost, for example, the budding of a particular plant or flower, or the first flight of migratory birds. Observers note the dates when birds or frogs lay their eggs or when bees build their hives. The following year they will compare the timing of the same events. Over the years these observations are used for forecasting purposes. People have been making such phenological observations since preagricultural times, a practice that has given rise to a great deal of weather folklore. Some of these sayings are forecasts, but others take the form of admonitions. "When the sloe tree is white as a sheet, sow your barley whether it be dry or wet" is a typical example.

St. Matthew brings on cold dew.

St. Matthew's Day is celebrated on the day before the autumn equinox in the Northern Hemisphere, which is usually September 21. With chillier weather comes the possibility of the first frosts—the cold dew of the saying. But there is no guarantee that the frost will not bide its time.

For every fog in August, there will be a snowfall in winter.

The belief that fog in August will forecast a snowy winter is so widespread that an Australian study discovered that it even had its adherents in Singapore, a part of the world not known to get any snow. Fog is produced when the humidity reaches 100 percent. This buildup of moisture might be considered a sign that a damp winter was in store. However, an unusually cool August, when it follows a hot July, is supposed to mean a cold but dry winter. It's hard to come by any scientific evidence to confirm that a foggy August has anything to do with the forthcoming winter, but that fog would correspond to snowfall is pure fantasy.

If St. Bartholomew's be clear, a prosperous autumn comes that year.

St. Bartholomew's Day falls on August 24. With the sun directly overhead, August is typically the hottest month of the year in the Northern Hemisphere. This British saying predicts that the days and weeks ahead will be warm, clear, and sunny. The weather is supposed to be auspicious for harvesting and fruit picking, too. But as is often the case, there is just as good a chance that autumn will prove disappointing.

Winter is six weeks ahead when the frost aster blooms.

This saying, popular in the American Midwest and Great Plains, especially Kansas, is based on a botanical confusion. Many people mistake the annual daisy fleabane for the perennial frost flower. Both daisy fleabanes and frost flowers are wildflowers, and can look similar. However, fleabanes have already bloomed and have begun to die by the time the frost asters begin to flower. If people believe that the former are the latter, they might be prompted to fear that an early winter really is coming. Moreover, there are many different types of frost flowers, and they go by many names—frost aster, hairy aster, and white oldfield aster. "The closer a bloom period is to a first frost, the more likely you are to think the two events are related—which, given short-term weather trends, could have a bit of validity," noted a Kansas researcher. "The tie would be sort of like the predictive value of early bird migrations, which do signal colder weather to the north. After all, that colder weather could move south… or not."

THE SEASONS

Seasons occur on Earth because of its location in space. We know that Earth rotates, or spins, while orbiting the sun, following an elliptical (oval-shaped) path. The plane of Earth's orbital path around the sun is known as the ecliptic. Earth is also tilted on its axis of rotation. (The tilt varies from about 22 degrees to 24.5 degrees.) With Earth tilted to one side, the angle of the sun's rays will strike Earth with varying degrees of intensity, depending on the time of year. If Earth was not tilted, sunlight would be most pronounced at the equator—that is the angle of incidence—and we would not have seasons. If the Northern Hemisphere is tilted toward the sun during Earth's orbit, it will receive the most direct sunlight. (In scientific terms, the angle of incidence is higher.) By contrast, the Southern Hemisphere receives less direct sunlight. So it follows that temperatures will be warmer in the Northern Hemisphere and cooler in the Southern Hemisphere. More simply put, it is summer north of the equator and winter south of it.

In the winter in the Northern Hemisphere, the sun will rise in the southeast and then, as the days go on, it will begin to edge slowly to the north, so that by the first day of spring—the spring or vernal equinox—the sun will rise directly in the east. The sun continues on its northerly path until the first day of summer, the summer solstice, when it will rise in the northeast. It then changes course and begins a steady advance to the south, rising directly in the east on the first day of fall—the autumn equinox—before resuming its southerly journey that culminates on December 21—the winter solstice. Then the cycle begins all over again.

Six months later, in December, the South Pole is more exposed to direct sunlight and the seasons are reversed. The position of the sun in relationship to Earth has a direct impact on the weather. The higher in the sky the sun is, the more direct the sunlight and the warmer the temperatures (which is what happens in summer). However, when the sun is lower in the sky, sunlight reaches Earth at an oblique angle—its light has to travel through more atmosphere—and temperatures are lower and the colder temperatures of winter prevail.

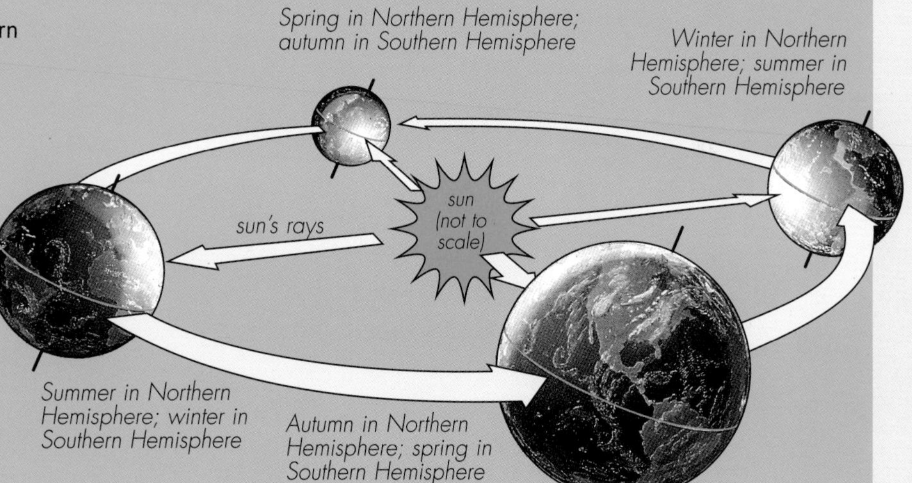

Jesus and the Fishermen
Jesus is frequently described as a fisherman (second only to his characterization as a shepherd). Many of his disciples were actual fishermen, which meant that they had to be aware of weather conditions.

WEATHER AND THE BIBLE

References to weather prediction abound in both the Old and New Testaments. For the most part, they are instructive or cautionary, rather than serving as mere description. The passages generally serve as reminders of God's power. One notable exception occurs in Matthew 16:2–3, when Jesus instructs a group of fishermen: *"When it is evening, you say, 'It will be fair weather, for the sky is red.' And in the morning, 'It will be stormy today, for the sky is red and threatening.'"*

More typical is the following verse from Psalm 107:25, 29: *"For He commandeth, and raiseth the stormy wind, which lifteth up the waves thereof.... He maketh the storm a calm, so that the waves thereof are still."*

Because winds are invisible yet powerful, they make for a convenient expression of God's will on Earth:

- Exodus 10:12–13: *"... And the Lord brought an east wind...[and] the east wind brought the locusts."*
- Exodus 14:21: *"... And the Lord caused the sea to go back by a strong east wind ...."*
- Exodus 15:10: *"Thou didst blow with thy wind, the sea covered them: they sank as lead in the mighty waters."*
- Genesis 8:1–2: *"... And God made a wind to pass over the earth, and the waters assuaged...and the rain from heaven was restrained."*

But the Lord's wrath can also express itself through the weather, and it does so frequently:

- Psalm 78:47: *"He destroyed ... their sycamore trees with frost...."*
- Psalm 18:12–14: *"... He shot out lightning, and discomfited them."*
- 1 Kings 8:35–36 (2 Chronicles 6:26–27): *"When heaven is shut up, and there is no rain, because they have sinned against thee; if they pray...and turn from their sin, when thou afflictest them: Then hear thou ... and give rain ..."*
- Psalm 78:47–48: *"He destroyed their vines with hail.... He gave up their cattle also to the hail...."*

Much rain in October, much wind in December.

October can be wet, and in certain parts of the world it marks the start of the rainy season. In Israel and the Palestinian Authority, for example, the rains begin in late October or early November and persist for the next two months; they are followed by the heavy "winter rains" that last through March. These two rainy periods are cited many times in the Old Testament, confirmation that weather patterns remain much the same as they were in the third millennium B.C. In Europe, October can be gray and rainy, although there is usually one short spell of favorable weather around the 18th of the month. This spell of dry, sunny weather is called St. Luke's Little Summer because the feast of St. Luke (who was a physician) takes place on this day. Southern England basks in the Little Summer, and so does Venice, whose inhabitants have a saying, "Pumpkins go stale on St. Luke's," presumably because the pumpkins receive too little moisture before being picked. However, there is little evidence that rain in October can be used to predict a windy December.

March comes in like a lion and goes out like a lamb.

This well known saying is derived from the observation that March begins in winter and ends in spring. In northern latitudes temperatures are generally higher by the end of the month than during its first weeks. Moreover, March typically announces itself with strong winds that tend to diminish as the days wear on. We may have to look to the heavens to determine why March is associated with both a lion and a lamb; it turns out that the constellation of Leo, the lion, dominates the skies at the beginning of the month and the constellation Aries, the ram or lamb, prevails as the month winds down.

Better a wolf in the fold than a fine February.

Mild weather at the wrong time of year generally bodes no good, as this Welsh saying makes clear. The warmth will accelerate the growth of crops and mislead flowers into blooming prematurely. The warmer weather is deceptive, and when inevitable frosts arrive, they can do considerable damage to crops, causing grievous losses for gardeners and farmers.

WINTER FOLKLORE

There are arguably more proverbs, sayings, and legends associated with winter than with any other season. Not only does winter bring cold, ice, snow, and other forms of inclement weather requiring advance preparation, but conditions in winter will also have a great deal of influence over the planting and harvesting seasons to come. Without sufficient precipitation in winter, the ground will be less suitable for growing crops—and it will not make ski resort operators happy, either.

Squirrels

Squirrels are indigenous to Europe, Asia, and the Americas; related species are found in Africa. In addition to the familiar bushy-tailed tree squirrels, members of the genera Sciurus *and* Tamiasciurus *also include flying squirrels and ground squirrels such as chipmunks, woodchucks, and prairie dogs.*

Northern lights bring cold weather with them.

Scientists dismiss the possibility that the northern lights could possibly bring weather of any kind with them. That is because the northern lights occur in the thermosphere, an atmospheric layer that begins about 50 miles (80 km) above the Earth's surface. Nearly all weather phenomena occur in the troposphere, which extends from the Earth's surface up to about 11 miles (18 km) and are heavily influenced by factors such as jet streams and air masses. Moreover, the northern lights also occur during summer, and they have been observed both in the Arctic and in Mexico at the same time. That would mean there is no connection between auroras and any type of weather. However, the frequent correspondence between the appearance of the northern lights and an inflow of arctic air in the northern regions has convinced some observers of a correlation.

It will be a cold, snowy winter if squirrels accumulate huge stores of nuts.

Squirrels do not necessarily sense whether winter will be colder than normal, but researchers have found that at least for red squirrels in the Canadian Yukon and Europe, they can predict when the trees will give them more food. They use that knowledge to produce two consecutive litters just at the right time to take advantage of the unusual abundance. In most years, trees produce a small amount of seeds, which curtails the population of animals that rely on them for food. But with fewer seed-eaters, more seed is allowed to grow, a phenomenon known as masting. That means that there will be a larger amount of seeds in certain years.

An icy wind from the northeast: Snow will fall on man and beast.

This rhyme has a particular application in southeast England. Any snow that this area receives during winter usually comes from northeasterly winds that have picked up moisture from the North Sea. By the time those winds reach England, the water vapor has crystallized and falls as snow, mostly along the east coast.

If the ears of corn are plump, a cold winter will follow.

Employing corn as a weather indicator is reflected in a Pennsylvanian saying: "When the corn wears a heavy coat, so must you." This is mistaking cause for effect. A thicker corn husk is the result of a wetter, warmer summer, but it is not a predictor of the severity of winter. That this belief exercises some power probably says more about human psychology: People tend to think that the pendulum will swing back from one extreme to the other so that a hot summer is likely to usher in a colder winter. This is not to say that weather patterns in one season cannot have a significant impact on weather conditions. How much color the leaves will have in fall and how long those colors will persist will depend on average temperature, light, and water supply during the summer. For example, several nights of temperatures just above freezing will produce more brilliant red maple leaves, while a great deal of rain will dull the colors.

It will be a hard winter if smoke from the chimney flows toward or settles on the ground.

Smoke does respond to weather conditions. High pressure is associated with storms. As a high-pressure system (such as a cold front) approaches, the air becomes denser and sinks to the ground, preventing any smoke from rising very high. Smoke particles also tend to absorb moisture from the air. So if there is more moisture present in the air, the smoke will absorb it, and it will get heavier. Heavy, moisture-laden smoke will not disperse as easily as smoke with lighter, drier particles.

THE BLUE HAG OF SCOTLAND

Known as the Crone Goddess or *Cailleac Bhuer*, the Blue Hag was depicted carrying a staff made of holly topped with the head of a carrion crow. She is associated with winter, and is said to dwell in a land where it was perpetually winter. Her staff was rumored to freeze the ground wherever it was tapped and to bring death to anyone it touched.

Appropriately for such a fearsome creature, the Blue Hag was reborn on every All Hallow's Eve (Halloween). Her reappearance brought winter and snowstorms. The early Christian Church, fearing the pagan religion of the Celts, might have had an interest in making her out to be a witch with fearsome powers. In her original incarnation, she was known to guard animals in winter.

FARMER'S ALMANAC

The legendary *Farmer's Almanac* was founded in 1792 and claims to be "the oldest continuously published periodical" in the United States, with a circulation of 5 million. It carries features on the home, gardening, history, astronomy, and food; but it is best known for its weather forecasts that the publisher boasts are 80 percent accurate.

The accuracy of the forecasts is derived from "a secret formula based on sunspots, position of planets, and tidal action caused by the moon," even though they're prepared two years in advance.

When the wind is in this quarter [south southwest] at Martinmas, it keeps mainly to the same point right on Old Candlemas Day, and we shall have a mild winter up to then and no snow to speak of.

This English saying refers to two holy days in the Christian calendar: St. Martin's Day, on November 11, and Candlemas, which falls in February. St. Martin's Day was commemorated with a big feast (Martinmas), marking "the great winter killing" in France of oxen and cows (which were not slaughtered until cold weather set in). Candlemas (the feast of the Purification of the Virgin) was celebrated on February 2 in the Gregorian calendar. (The association with Groundhog Day is not coincidental.) Candlemas (so named because it was celebrated by the lighting of candles) incorporates pagan rituals that took place at the astronomical midpoint, or on the full moon closest to the first spring thaw. In the British Isles, good weather at Candlemas foretells severe winter weather. It is also the date that bears and wolves are said to emerge from hibernation to see whether it is safe to leave their lairs or whether they should burrow down for another 40 days. Traditionally, sailors were loath to set sail on Candlemas Day because of the likelihood of running into storms.

A year of snow, a year of plenty.

This saying is based on observations that a thick and persistent snow cover serves an important protective function. The snow will delay plant growth on farmland and the blossoming of fruit trees long enough to cushion them from the effects of killing frosts that can come as late as March and April. A continuous snow cover also prevents a cycle of thawing and freezing, which can destroy wheat and other winter grains. A generous accumulation of snow also means more runoff, ensuring a dependable supply of water for farmland in the spring.

It's spring and then suddenly it's winter.

Folklore of the eastern United States cautions gardeners to wait until after the dogwood trees have bloomed before planting. "Dogwood winter" typically occurs in May, catching people off guard after they've enjoyed an extended midspring warm spell. It is characterized by several days of cold, nasty weather, often with frost and sometimes snow flurries. Its unwelcome arrival coincides with the blooming of dogwood trees, which grow in habitats ranging from Texas to Minnesota eastward to the Atlantic shoreline, from Florida to southern New England. In regions where dogwoods are uncommon, the inclement interval is sometimes called "locust winter" or "blackberry winter." (Blackberry bushes blossom around the same time.)

In some areas it is known as "linsey-woolsey britches winter" because it is the last time when winter clothing made out of homespun linen and wool underwear needs to be worn. These spells represent a resurgence of a continental polar air mass after maritime tropical air masses have begun to dominate. Dogwood winter is regarded as a weather "singularity"—along with other anomalous events in North America such as "Indian summer" (a period of warmth after the first frost) and the "January thaw."

Weather Balloon Firsts

The first experiments in high-flying weather recording occured in 1749, when British scientists attached thermometers to paper kites to monitor the atmospheric temperature up to a height of 1,000 feet (305 m). The French scientist Leon Teisserenc de Bort carried out the first tests with balloons in 1896. The first flight test with a radiosonde took place in 1931.

WEATHER BALLOONS

Weather balloons are used to track temperature, pressure, relative humidity, and wind speed and direction in the upper atmosphere, monitoring regions that are often inaccessible to other types of technology, such as ground radar, aircraft surveillance, and weather satellites. About 2,000 balloons are launched around the world every day. Most are filled with helium, although some rely on hydrogen. Pilot balloons are used to obtain data on wind speed and direction of winds at different altitudes. Smaller altitude limit balloons are used to determine the altitude of cloud bases. The workhorse of weather balloons is a larger, teardrop-shaped balloon equipped with a radiosonde—a package of instruments weighing 1 pound (almost half a kilogram). The radiosonde sends to the weather service a constant stream of data relating to atmospheric conditions—1,000 to 1,500 readings—for the duration of the flight. The balloons take the package aloft to an altitude of 90,000 feet (27.4 km) or above. If temperatures drop below -130°F (-90°C), the balloon's diameter swells and the balloon tends to pop because of low air pressure. At that point the radiosonde descends to the ground, carried by a parachute to ensure a safe landing. The package also contains mailing instructions so that anyone who discovers it can return it to the weather service.

SPRING FOLKLORE

NATIVE LEGEND
The Cherokees tell of an era when they were "new upon the Earth," and decided that they could do without night. The Creator Ouga granted their request, but with the cessation of night the forest began to grow so thick that it became impenetrable. Realizing their mistake they appealed to Ouga for perpetual night. But endless night was even worse; crops wouldn't grow and no one could hunt. The people began to perish from cold and hunger. Understanding the error of their ways, they implored Ouga to restore day and night, and the weather was once again pleasant.

Snowdrops
These spring flowers may bloom prematurely in unseasonably warm weather, heralding a false spring.

Spring is welcome because of the promise of warmth and longer days, but spring is also untrustworthy because it cruelly flirts with our expectations, taking back with one hand what it just gave with the other. So folklore is full of warnings against investing too much confidence in warm days early in the spring. Tomorrow, after all, might always mark a brief but painful reprise of winter.

When the first robins appear, spring is at hand.

There is a distinction to be made between spring as indicated on the calendar (usually March 21, the vernal equinox) and the onset of springlike days, as the American author and educator Henry van Dyke observed in *Fisherman's Luck:* "The first day of spring is one thing, and the first spring day is another. The difference between them is sometimes as great as a month." Inhabitants of the Hudson Valley in New York who have long memories refer to early spring snowfalls as "robin's snows," an implied rebuke to the bird traditionally considered a herald of the season.

Spring can't come until the spring peepers have been shut up twice.

Spring peepers refer to frogs, most likely tree frogs (or spring frogs). Admirers have characterized the mating calls of the males on early spring evenings as almost symphonic. (Others have not been so charitable.) This saying suggests that spring cannot be said to have come until at least two cold snaps have quieted the frogs.

April sun and April showers bring forth summer flowers.

Earth and its atmosphere in the Northern Hemisphere warm up rapidly after the winter. When the atmosphere becomes unstable and temperatures in the atmosphere differ sharply from those on the ground, thunderstorms and "April showers" can develop. In Wales, where this saying originates, as well as in other parts of Europe and in North America, the combination of rising temperatures and showers is ideal for the early growth of spring flowers such as snowdrops, wild violets, and wood sorrel. The position of the jet stream in April also plays a role in promoting showers.

The jet stream, a band of very strong winds at around 30,000 feet (9.14 km) above the surface of the Earth, exerts a great deal of influence over weather patterns. High- and low-pressure systems form, strengthen, or weaken as the jet stream accelerates or slows. As the jet stream starts to move northward in early spring, it allows a depression to form. Depressions are systems that can generate severe weather—hurricanes and tornadoes, for instance—and frequently culminate in the Atlantic. These systems can bring bands of rain or snow showers, and blustery winds to Ireland and Scotland or to the eastern seaboard of the United States. As a result, April days can begin under sunny skies and end in a downpour.

Who shears his sheep before St. Servatius' Day loves his wool more than his sheep.

St. Servatius' Day occurs on May 13. Even though the skies in April and early May are becoming clearer and temperatures are higher, nights may be chilly, putting a shorn sheep's life at risk.

Shivering Sheep!

The St. Servatius' Day saying contains a note of caution: Don't trust the weather! Spring is a fickle season.

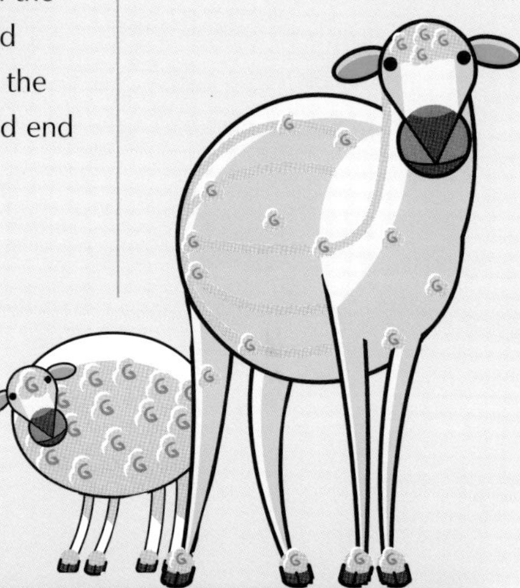

WEATHER RADAR

Radar sends out beams of energy, and as they meet objects, the objects reflect the beams back to the radar. Two factors are assessed: How much time does it take for the beam to return, and how much of the beam does come back. These objects can include rain, snow, sleet, and even insects. Objects of no interest to meteorologists are called "ground clutter." If a large portion of a beam is returned, then the object is referred to as one with high reflectivity, which is expressed by brighter colors. Conversely, objects that send back only a small portion of the beam have low reflectivity and appear as darker colors. The Doppler radars referred to in weather reports can also measure wind speed and direction. Reflectivity depends on both the intensity and the type of precipitation. Because hail and sleet are made of ice, they are very reflective, while snow can scatter the beam, resulting in deceptive images that make it appear as if the snowfall is light when it is the opposite.

WEATHER MAP SYMBOLS

Deciphering symbols on weather maps is actually easier than it may look at first impression. Once you understand their meaning, you can easily track various weather conditions, such as wind speed, the extent and type of precipitation, the movement of storm systems, and the approach of warm and cold fronts. Meteorologists around the world use the same symbols so that they can exchange information easily. The meaning of some of these symbols is not obvious to the public, so symbols commonly used on television and newspaper weather maps tend to be simpler. Meteorologists measure windspeed in knots: 1 knot is 1 m.p.h. (1.85 kph).

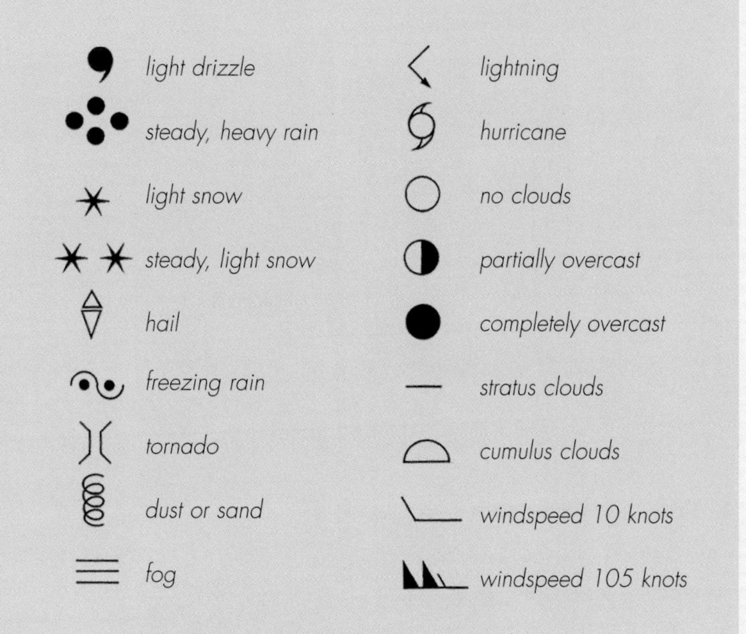

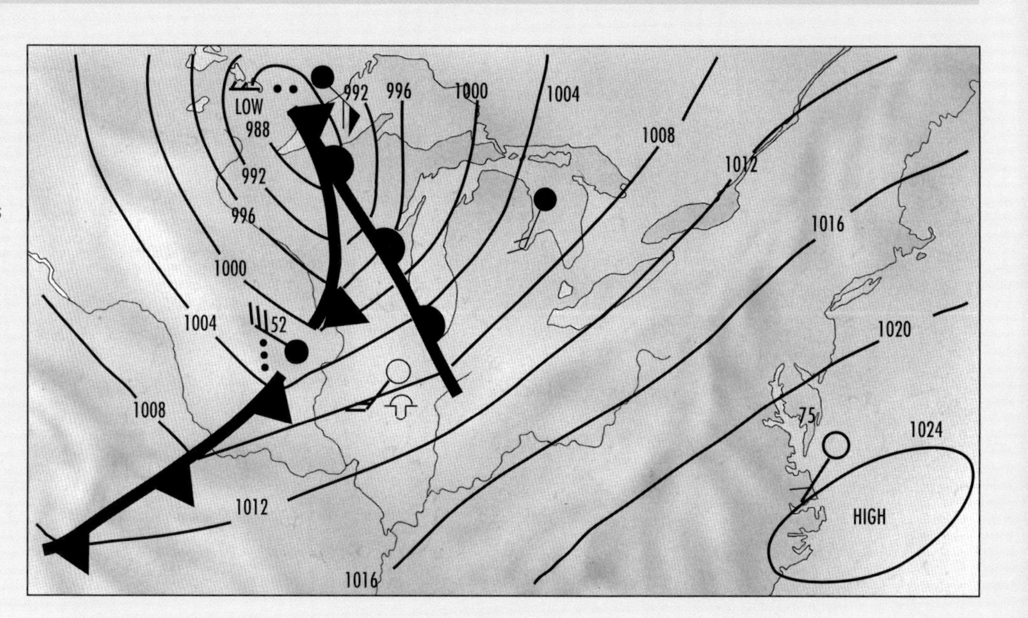

Weather Maps

Isobars provide information about air pressure and wind speed and direction. The isobars will be closer together in low-pressure systems, and farther apart in high-pressure systems. The closer the isobars, the stronger the winds in that region. Cold fronts are identified by barbed lines, warm fronts by semicircles.

If the goose honks high, fair weather.

This saying applies in general to flocks of birds whose migrations tend to be daily and over short distances. That explains why the sight of a lone crow is not considered favorable. Migrating birds fly higher in fair weather than in foul. Wild geese flying north are a sign of warm weather. The higher they fly, the more settled the weather. Birds fly as high as they can while they're migrating in order to make use of their ceiling (or altitude limit) to increase their speed.

The ceiling is lifted in good weather, when high-pressure systems prevail, and lowered in low pressure, when the air is more turbulent. The birds are aided in their flight by rising packets of warm air from the ground, called thermals. The early arrival of migratory birds such as swallows and martins promises dry, warm conditions throughout their nesting seasons in late April, May, and June. Employing nonmigratory, solitary species such as thrushes, wood pigeons, and crows for forecasting purposes was more problematic. It required more work to investigate their nesting season, if you wished to discover whether they were building their nests higher or lower. Higher nests supposedly meant that there was a greater probability of a warm and dry summer.

Goose in Folklore
In early Christian tradition the goose was thought to presage the coming of the winter solstice—known as "The Great Freezing."

A wet spring, a dry harvest. Spring rain damps, autumn rain soaks.

This saying comes from Russia. Generally speaking, the cooler it is in spring, the better the crop. For some crops such as wheat, hot weather is damaging. That's why farmers subscribe to the adage "early plantings beat the heat." At the same time, rushing to plant too early risks exposing the immature plants to a killing frost. However, rainy weather early in spring can also make it impossible for farmers to plant at all. This results in a shorter planting season, which can also have an adverse impact on the crops. It isn't simply the amount of rainfall that affects the crop; it is the way in which the rain comes down. Downpours cause crusting on the surface of the soil; that allows water to stagnate. Standing water can kill seeds. Heavy rains at harvest time can cause the runoff of fertilizers, insecticides, and herbicides as well, which not only leaves plants unprotected against pests, but also has the potential to pollute nearby rivers.

ATMOSPHERIC PRESSURE

Atmospheric pressure refers to the weight of air at a particular location. Atmospheric pressure is greater at sea level than it is at the top of a mountain. Increases and decreases in air pressure play a crucial part in determining the prevailing weather.

High atmospheric pressure is often linked with clearing conditions, while an area of low atmospheric pressure is associated with stormy conditions.

SUMMER FOLKLORE

Summer is a season of special interest in folklore because it occurs between the planting season of spring and the harvest in fall. If the weather took a turn for the worse in the summer, crops could be ruined—and so could the rest of the year as a result. Many festivals were focused around the summer solstice, when bonfires would be lit to purify the land and protect crops from blight and animals from the threat of diseases endemic to hot weather.

Barnaby bright,
the longest day
and shortest night.

St. Barnabus' Day falls on June 11, whereas summer begins officially on the summer solstice (on or around June 21), when the day is the longest of the year and night is shortest. The saying is not inaccurate so much as it is anachronistic. The saying dates from a time when the Julian calendar (introduced by Julius Caesar in 46 B.C.) was used. It has since been replaced by the Gregorian calendar that we rely on today, which accounts for the 10-day difference. The difference arose because the Julian calendar was out of sync with the tropical year (the interval between vernal equinoxes), a discrepancy resolved by Pope Gregory XIII (for whom our current calendar is named), which resulted in the solstice being relocated. The solstice still moves, but only slightly; it will change by one day in 3,000 years.

Rain on Midsummer's Day or the Feast of the Nativity of John the Baptist (June 24) means a wet harvest.

Midsummer was celebrated around the time of the solstice, since antiquity. The popularity of these traditions alarmed early Christians, who subsequently incorporated such celebrations into their own liturgical calendar. That was why St. John's Day was adopted, since it conveniently fell close to the solstice (in the Gregorian calendar). Historians have noted that ordinarily the liturgical calendar would commemorate the day of a saint's martyrdom (August 29), not his nativity (which is what is marked by St. John's Day), underscoring the need the early

DOG DAYS

The Dog Days, a term first used by the ancient Romans, derive their name from the rising of Sirius (the Dog Star), the brightest star in the sky. (*Canis* is Latin for "dog.") It refers to the hottest days of the year in the Northern Hemisphere, from about July 3 to August 11. There is no equivalent period in the Southern Hemisphere.

They result from a lingering maritime tropical air mass, which brings with it hot temperatures and thunderstorms. However, the Dog Days can be pleasant because of the influence of the jet stream. In that event the system tends to persist into the autumn and even into the winter, as it does in England and Wales. That accounts for the Welsh saying:

"Dog Days bright and clear,
indicate a happy year."

Christian fathers felt to impose their stamp on what previously had been a pagan rite. Wet harvest seasons are bad news; they result in low yields and poor qualities in such crops as corn, cotton, and soybeans. Rainy conditions also force farmers to hasten the harvest before even more crops are ruined; but on large farms that use combines, one consequence of rushing the harvest is field rutting. The rutting comes from the treads of the machines that dig deep into the muddy earth. Ruts create rough, uneven areas that often retain water. Even though it may rain in late June, though, there is no reason to believe that late summer will be especially rainy. The prospect of a rainy summer does increase if rain becomes more frequent in July or August, which are typically drier and warmer than any other months.

THE WATER CYCLE

All life is more or less dependent on the water cycle, which is based on three processes: evaporation, condensation, and precipitation. In one form or another, water is continuously circulating between the Earth and its surrounding atmosphere. For example, water that evaporates from the oceans or the ground is sucked up into the air. The amount of water vapor in the air is known as humidity. Ordinarily, air is capable of holding only a limited amount of water; but as the temperature rises, air can accommodate far more water—the quantity doubles for each temperature increase of 18°F (10°C), which is why humidity is especially intense on sweltering summer days. As temperatures cool, the excess water vapor in the air condenses into liquid droplets or ice crystals, forming either clouds or fog. Clouds, of course, can assume many forms, and owe their appearance to the motion of air that created them. When the droplets and crystals in clouds grow large enough, however, the clouds are no longer able to accommodate them and they fall to the ground as precipitation. Depending on the temperature in the air and on the ground, precipitation can descend as rain, drizzle, freezing rain, snow, hail, ice pellets, or sleet.

Morning Clouds

The appearance early on a summer day of different types of clouds, especially cumulus and cumulonimbus, may indicate that thunderstorms will follow later in the day. Cumulus clouds may form cumulonimbus clouds, which are capable of growing into the higher reaches of the atmosphere where the water vapor freezes, a process that can lead to the precipitation and turbulence seen in severe thunderstorms.

When the swallow's nest is high, summer is dry. When the swallow's nest is low, you can safely reap and sow.

For the most part, swallows choose higher perches to build their nests—cliff faces, or in caves if available—but they also favor accessible buildings, or build their nests under bridges. Female barn swallows typically lay four or five eggs. There is no evidence that their nesting habits will predict either weather or harvests. However, it should be noted that their numbers have declined in parts of Europe and North America because agricultural development has reduced the supply of insects that swallows rely on for food. In Estonia, the swallow is the national bird. According to local legend, anyone killing a barn swallow will turn blind.

August 17: Cat nights begin.

"Cat nights" refers to a period in Europe when belief in witchcraft was widespread. According to old Irish legends, a witch could turn herself into a black cat eight times with impunity, but on the ninth time she couldn't change back. Supposedly, this is the origin of the expression "a cat has nine lives." Cat nights occur in late summer when the constellation of Leo is more prominent. Among the most ancient of known constellations, Leo is depicted as a six-star, curving sicklelike configuration representing the lion's head. There are two other minor constellations in which members of the cat family can appear close together: Leo Minor (the smaller lion) and the Lynx.

St. Bartholomew brings the cold dew.

St. Bartholomew's Day falls on August 24; some weather lore designated this day as the beginning of autumn. In fact, autumn doesn't begin until September 21, the autumn equinox. However, people endowed St. Bartholomew's Day with a certain predictive power since it was construed as an indicator of what autumn would be like. The term "cold dew" refers to frost. Dew and frost form in different ways, although the two phenomena are related. Dew is produced when the air is cooled sufficiently during the night to condense on surfaces. The temperature at which the air can no longer hold all the water it contains is called the dew point. The dew usually evaporates once the sun rises. Frost consists of ice crystals formed by the deposition of water vapor on a relatively cold surface of an object.

The more cloud types at dawn, the greater the chance of rain.

This saying describes a situation that typically occurs only in summer, when cumulonimbus clouds are present with other types of clouds, especially cirrus. The variety of clouds was seen as a sign that thunderstorms were likely later in the day. Cumulonimbus clouds, formed from cumulus clouds, are tall and dense, and develop alone or in clusters or array themselves along a cold front in a squall line. These clouds often form near the oceans, where sea breezes provide the fuel they need to push them higher into the atmosphere.

For these clouds to develop, they require a good deal of moisture; a mass of warm, stable air; and a source of energy to lift packets of warm, moist air higher into the atmosphere, as cooler air rushes in to fill the void closer to the ground. Once the water vapor rises high enough, it begins to cool and condense into water droplets. It takes energy to do this, which creates latent heat in the atmosphere surrounding the rising packet of air. That heat creates ever more energy, which continues the lifting process. As the air rises farther, the water droplets continue to cool; but since the temperatures are even cooler, ice crystals begin to form. The ice crystals become too heavy to remain suspended in the clouds and, pulled by gravity, tumble down through the atmosphere in the form of hail. The downward movement creates a violent downdraft even as air is rising from the ground, producing an updraft. All this turbulent activity causes static electrical charges to build up within the cumulonimbus cloud. The electricity generates the thunder (and lightning) that gives thunderstorms their name.

THE BASQUE WEATHER GODDESS

The Basque goddess Anbotoko Mari is believed to have a special association with the weather. When she would leave her cave, storms or droughts would occur. When she and her consort Majue traveled together, hail would fall. (The couple had two children: one benign and one evil spirit.) When Mari stayed put in her cave, it was usually raining. Mari was regarded as the personification of Earth, who held influence over the forces of nature, particularly thunder and wind. In spite of all the stormy weather that her appearances would produce, Mari was actually worshipped as a protector of travelers and herdsmen. She is generally depicted riding through the sky on a chariot drawn by horses or rams.

AUTUMN FOLKLORE

Autumn, like spring, is a transitional season, taking its weather from both summer and winter; it is a time when observers have traditionally closely monitored natural phenomena, looking for indications as to the severity of the winter ahead. Squirrel behavior, first frosts, and the time when trees lost their leaves were all subjected to unusual scrutiny. If it was going to be especially cold, food would be stored up.

If St. Michael brings many acorns,
Christmas will cover the fields in snow.

St. Michael's Day falls on September 29, which is about the time that acorns begin to fall from the oak trees in the Northern Hemisphere. Native Americans, like Europeans, were pleased when a good acorn crop fell because acorns were used for soups, bread, and pancakes. But if the acorn crop was too bountiful, they feared that it meant that a cold snowy winter was imminent. Researchers who have studied large acorn crops of three oak species in central Missouri have concluded that this belief is not warranted by the facts. Within a species, large crop yields of individual oaks all occurred in the same year, but different species produced large crops in different years.

The size of a given acorn crop was determined by both flower abundance and whether the flowers survived long enough to bear fruit. Weather during the previous spring and summer had a significant impact on the crop; so did previous acorn production. But each species showed its own distinctive pattern: Red oak acorn yields differed more from red oak acorn yields in the previous three years than the black oak yields did. The researchers reasoned that the oaks must store up resources over time in order to produce a large crop. Moreover, species have different reproductive cycles—black oak has two-year cycles; white oak, three; and red oak, four. So reproductive events, along with weather, must be taken into account as to how large an acorn crop will be produced each year from a different species of oak. Taking all these factors into account, weather does play a major part in determining the size of acorn crops in autumn, but the yields in no way predict what the approaching winter will be like.

FALL FACTS

- Deciduous trees and plants lose their leaves in autumn. Evergreen trees will lose their leaves continually. This occurs mostly in spring when they turn yellow, but never all at once.
- In autumn, the leaves of certain trees, including birches, tulip poplars, redbud, and hickory, always turn yellow. Leaves of sugar maple, dogwood, sourwood, sweet gum, and black gum usually turn red, but occasionally yellow.
- The colors of leaves serve no biological function, unlike the colors of flowers, which attract pollinators.
- The expression of red pigment in leaves depends on abundant sunlight; otherwise the leaves will turn yellow.
- Of the 650 species of bird that nest in the United States, about 520 migrate south for the winter during fall.

AIR MASSES

An air mass is a body of air that has similar temperatures and humidity throughout. Such masses can extend over several thousand square miles. They originate over large flat areas in which the air has become sufficiently stagnant that it takes on the characteristics of the underlying land. They circulate around the planet and sometimes cause dramatic weather changes. The collision of these masses can also produce very dramatic weather conditions—for example, when an arctic body of air meets a warmer air mass. Air masses are categorized by the region over which they formed. The source regions can have one of four temperature attributes: equatorial, tropical, polar, or arctic. They can also have two moisture characteristics—continental or maritime—meaning that they originated over either land or sea.

Air masses are classified in five ways:

- *Continental Arctic* or *Antarctic* (cA), with extremely cold temperatures and very little moisture
- *Continental Polar* (cP), with cold temperatures and dry conditions
- *Maritime Polar* (mP), with cool temperatures and moist conditions
- *Maritime Tropical* (mT), with warm temperatures and plentiful moisture
- *Continental Tropical* (cT), with hot, dry conditions

On some weather maps, a "K" or "W" is attached to the two-letter designation of an air mass. The "K" means that the air moving across a region is colder than the land surface temperature, and "W" means that the air is warmer than the land surface temperature.

Wet air masses are formed over the oceans; dry air masses are formed over continents. Equatorial air masses are all considered to be wet, since much of the equatorial zone is covered by rainforests that can add as much moisture to the air as the equatorial oceans. Arctic or Antarctic air masses are considered dry because there is little evaporation derived from the polar oceans. Their temperatures are so cold that even at saturation, the absolute humidity is very low. When an air mass moves out of its source region, it undergoes changes of temperature and humidity as it passes over areas with different characteristics. Polar air masses formed over the Arctic, for example, will drift southward, where temperatures are warmer, so that the air masses will warm as well.

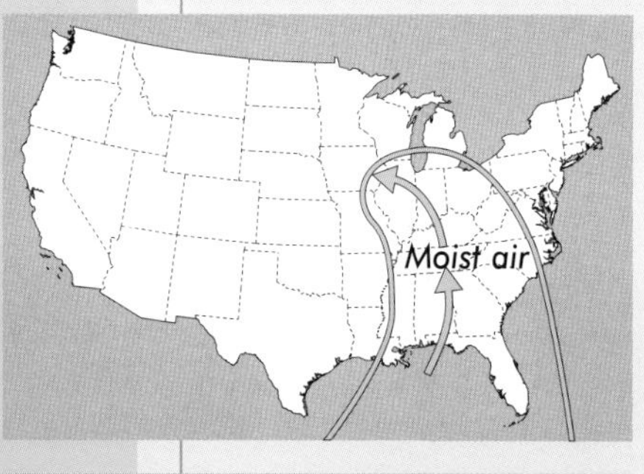

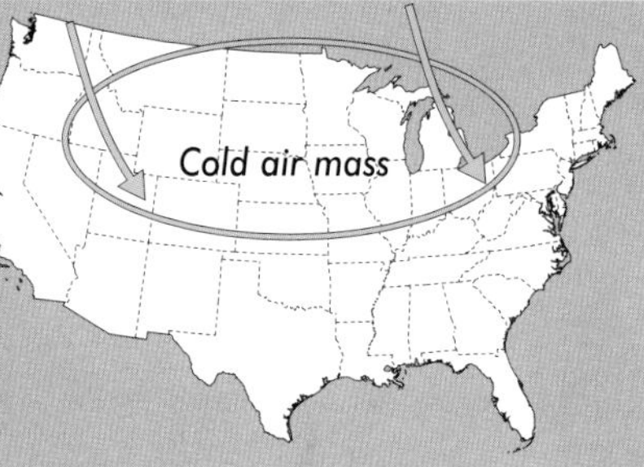

Seasonal Variations

When maritime tropical air masses from the Caribbean and Gulf of Mexico move into the southern United States, they can be accompanied by intense precipitation. This region is prone to tropical storms from June to November (top). In winter, continental polar air masses creep down into the United States from Canada. If conditions are right, these air masses can sit for several days, bringing colder temperatures, blustery winds, and snow (above).

Autumn Leaves

New research is revealing why leaves of certain trees are more vibrantly colored than others in the autumn. Red pigments—called anthocyanins—in maples, oaks, and dogwoods act as sunscreens, shading sensitive photosynthetic tissue. This allows time for trees to reabsorb vital nutrients from their leaves before they fall.

If the leaves do not fall by St. Martin's, expect a cold winter.

St. Martin's Day is celebrated on the closest Sunday to November 11. So leaves that are still hanging on trees would be very late in falling indeed. But do lingering leaves portend a colder winter than normal? The chemical process that takes place within leaves in autumn is triggered by the diminishing length of days. Or put another way: Leaves burst into color and die as a result of dwindling daylight and longer periods of darkness. Only deciduous trees are able to "sense" the fact that nights are getting longer. (Deciduous is derived from the Latin word meaning "to fall down.") Evergreens, on the other hand, will lose leaves, but gradually over time and never all at once. As the days grow colder, they bring an inflow of dry air. That's because cold air cannot hold as many water molecules as warm air. The tree can no longer afford to keep replenishing leaves with moisture since there is less and less of it all the time; so to preserve what water it does have, it gets rid of its leaves in a kind of triage. The process is unstoppable, but it can be delayed by a period of unseasonably warm days, in which case a tree will marshal what nutrients it has in reserve in response to the warmth, allowing the leaves a

CLIMATE CHANGE

Climate change is not new; changes in climate (or average weather) have occurred in cycles that can range from decades to millions of years. The cause of climate change may be natural—for instance, volcanic activity that obscures the sun with dust, producing cooler temperatures—or manmade. Since the Industrial Revolution in the nineteenth century, humans have contributed to the warming of the Earth's surface, perhaps as little as 32.7°F (0.4°C) or as great as 33.4°F (0.8°C), largely by the growing use of fossil fuels. This results in a buildup of greenhouse (carbon-based) gases in the atmosphere as well as in the oceans. These gases can act as an insulator, preventing heat from escaping into the upper atmosphere as they would ordinarily, leading to warmer temperatures. The warmest years of the twentieth century have all occurred in the last 15 years. A computer model predicts that by 2040, the arctic ice cap will be completely melted because of global warming. Melting ice sheets will cause the seas to rise, resulting in greater flooding and submerging some islands in the South Pacific altogether. Temperatures in some parts of the world might actually become colder because of the icier seas. More extreme weather events—droughts, hurricanes, intense snowstorms—are also being blamed on climate change. If nothing is done to curb energy production, the principal source of carbon emissions, scientists predict even more dramatic and severe climatic and meteorological changes in the future.

reprieve. If the weather turns suddenly cold earlier in the season, however, the process is aborted; the leaves don't turn color and don't fall off. Instead they simply die. That accounts for the desiccated brown leaves that cling to branches even in winter. It is believed that there may be an evolutionary purpose for a few leaves to remain in place in winter; by providing sustenance for hungry animals prowling the snow, the tree may be protecting its bark and branches from their depredations. That may explain the logic behind this saying from Wales. However, leaves stay on deciduous trees because of colder weather, not because their presence necessarily means that weather conditions will remain unusually cold.

***If All Saints (November 1) brings out water,
St. Martin's (November 11) brings out Indian summer.***

Indian summer can occur in October or in November in the United States and Canada. It is a warm spell that occurs in late fall after the first hard frost, or alternatively after freezing temperatures have become common. (Hard or killing frosts are those that damage fruit and vegetables.) Indian summer, like St. Luke's Little Summer, is a reprieve resulting from the return of a maritime tropical air mass. Neither interval lasts for very long.

***No warmth, no cheerfulness, no healthful ease,
No comfortable feel in any member,
No shade, no shine, no butterflies, no bees,
No fruits, no flowers, no leaves, no birds—No-vember.***

These lines were written by Thomas Hood (1799–1845), a British humorist and poet. November has a reputation for bleak weather and gray skies. In fact, in Europe and in North America, November can actually prove more benign than the verse makes it out to be. In parts of Europe, for example, the middle of the month is often more settled, with an anticyclone activity producing sunny days and frosty nights. Anticyclones are high-pressure systems characterized by sinking packets of warm air. These systems typically bring warm dry days and settled weather for sustained periods of time. The weather in November depends in part on the kind of summer and autumn weather that preceded it. In that respect, folklore is correct: Weather in one season can predict, to some extent, the weather in another season.

2

COLD SPELLS, SNOW, AND HAIL

The annals of folklore are full of signs of cold weather. Not surprisingly, it was believed that if squirrels buried acorns earlier than usual, or if there was a larger acorn crop than usual, a long, cold winter was in store. But a warm or thundery fall was also considered a sign of the arrival of bitterly cold weather. Mild winters were viewed with suspicion because it threatened to upset the natural order. Mild winters have become more the norm recently in many parts of the world, in part because of climate change.

COLD SPELLS

According to a Tyrolean legend, vineyards used to grow in parts of the Austrian Alps, and snow and frost never touched the ground; but one day snow began to fall, and farmers gathered their families to warn them that a thousand years of cold were coming. The fear of cold found in many folklore sayings reflected a grim reality: Winters were long, hard, and lethal. Between the years 1150 and 1460 and again between 1560 and 1850, Europe suffered through what became known as the Little Ice Age. Millions of people perished after the failed harvest of 1693, while 1816 became known as "the year without a summer."

When walls in cold weather begin to show dampness, the weather will change.

This is probably a good indicator that the humidity is increasing, which may give warning of cloudy or rainy weather and a warming trend. It also might indicate that you need to hire someone to inspect your house for leaks.

December cold with snow, good for rye.

Because it provides insulation, snow can protect seedlings and plants that, if exposed to frigid conditions or killing frosts, could suffer grave damage or perish altogether. However, a snow in December does not necessarily mean that snow will continue to remain on the ground or that there will be fresh snowfalls as winter goes on. A sudden thaw could undo any good that an early snowfall had done.

When there are three days cold, expect three days colder.

Whether a cluster of cold days will mean that even more bitterly cold weather follows depends on so many factors as to render this saying impractical as a forecasting guide. Such a saying may come from anecdotal evidence; once a weather pattern takes hold, whatever its nature, it tends to persist. The term "cold spell"—in the sense of a bout—has been in circulation since the late 1700s.

FREEZING FACTS

- There are actually two winters: meteorological winter and astronomical winter. The former is the period from December 5 to March 5—usually the coldest months in the Northern Hemisphere. Astronomical winter begins on the winter solstice and ends on the vernal equinox.
- The Mississippi River froze from its source to its mouth in New Orleans in February 1899. Ice was 2 inches (5 cm) thick at New Orleans. The only other similar freeze was in 1784.
- The coldest inhabited village on Earth is Oymyakon in Siberia, where temperatures may fall to -89°F (-67°C).

HUNGRY POLAR BEARS

Polar bears' main prey is seals, which they catch when the seals come up for air through holes in the Arctic ice. With rising global temperatures, the ice breaks earlier and freezes up later, causing the polar bear to starve for longer periods each year. Research shows that polar bears weigh on average 10 percent less than 20 years ago and produce fewer offspring each year, forcing a steady decline of the Arctic polar bear population.

A cold spell usually applies to wintry conditions. A cold snap, on the other hand, is more likely to refer to an unexpectedly cold period in spring when warm air is often supplanted by a polar air mass.

The harm that cold can do, especially to agriculture, cannot be underestimated. Cold can stunt tree growth, which will have a negative impact on forests. Extreme cold has the capacity to crack large trees and fruit, resulting in dehydration and death of the plants. Cold can dramatically diminish harvests, especially if there is insufficient snow cover. In severe cases cold is directly related to famine and consequent high mortality rates. Unsurprisingly, animals suffer, too: Insects disappear, and that in turn causes insect-feeding bird populations to fall. Freezing rivers and lakes can cause widespread death among fish and amphibians. The one positive thing that cold spells do is kill off large numbers of parasites.

With increasing evidence of the effects of climate change, a misapprehension has taken hold that cold spells will become less frequent. In fact, climate change may produce more frigid winters in some parts of the world. Melting ice from Greenland and the Arctic, which is already evident, is not only expected to lower the temperature of the Atlantic (while raising sea levels), but may shift the direction of ocean currents and bring much colder weather to affected regions.

In June 2006, to take one example of this counterintuitive finding, Australia reported record or near-record cold nights the second coldest June since 1950 because of the suspected influence of climate change. (June is a winter month in the Southern Hemisphere.) The particularly frigid cold spell was fueled by a high-pressure system that is typically associated with clear nights, low humidity, and light winds, conditions that prevail in fair weather, but which are also responsible for very cold temperatures.

THE FOLKLORIST WHO BEAT THE METEOROLOGISTS

The late weather observer Bill Foggitt of Yorkshire, England, based his forecasts on the behavior of sheep and moles, and phenomena such as the moistness of pinecones and seaweed to make his predictions. In 1985 the British Met Office, the national weather service, conceded that his forecast of an especially frigid winter made several years previously was correct.

In 1946 he had written a letter to a newspaper in which he stated that Britain could expect an arctic winter because of the untimely arrival in England of a flock of waxwings—exotic, crested birds that migrated from their home in Siberia. He was right.

Film Star Phil
In the film Groundhog Day, Bill Murray plays a weatherman who, sent to report the sighting of Punxsutawney Phil on Groundhog Day, is forced to live through the same day over and over until he mends his wayward ways.

If on February 2, the groundhog sees its shadow,
30 days of winter remain.
If not, spring will follow immediately.

The most celebrated day in weather lore is centered on an animal—a groundhog in the United States and Canada, but also a bear, a wolf, and a badger in Europe. The legend goes that if the animal sees his shadow, there will be six more weeks of winter; if he fails to see it, spring will follow soon.

In the Czech Republic, for example, the day is known as *Hromnice*, a day that has inspired several sayings such as "If the badger comes out of his hole on Hromnice, he'll be rushing back four weeks later." The occasion is marked in similar fashion to Groundhog Day, although sometimes it takes on more of a religious complexion. In many cultures February 2 is considered the day that will foretell whether spring weather will come soon or whether winter will persist for another six weeks instead.

This date is no coincidence, since it represents the approximate midpoint between the winter solstice and the spring equinox. This doesn't necessarily mean that the temperature is about to

GROUNDHOG DAY FESTIVALS

Groundhog Day, which is celebrated across the United States and Canada on February 2, has become a big moneymaker for the Chamber of Commerce of Punxsutawney, Pennsylvania, which boasts the most famous groundhog of all, Punxsutawney Phil. The town gained additional publicity with the release in 1993 of the movie *Groundhog Day*, starring Bill Murray as a befuddled meteorologist who wakes up every morning to find that it is Groundhog Day all over again. The annual festival dates from the late 1880s, owing to the efforts of a local newspaper editor and congressman. Groundhog Day events are also held in other states, including Nebraska, Tennessee, Georgia, Ohio, New York, Arkansas, North Carolina and California, as well as in Canada.

moderate, as anyone who has lived through February or early March in the Northern Hemisphere knows. Nonetheless, it still has meteorological significance insofar as it marks the end of a three-month period during which that part of the world receives the lowest amount of solar energy during the year. From early February on, the amount of solar energy will begin to increase. The history of this special day can be traced back to the Emperor Justinian I, who declared February 2 the Feast of the Purification of the Virgin in A.D. 542. (The day has roots in pagan ceremonies, too.) In medieval Europe this date became Candlemas Day, the date for the blessing of the sacred candles that were to be used in the coming year. Candlemas commemorates the occasion on which the Virgin Mary would have attended a ceremony of ritual purification under Jewish law. The Gospel of Luke states that shortly after Mary was purified, Jesus was presented in the temple of Jerusalem. That date, 40 days following the Nativity, fell on February 2.

Groundhog Day
The seer of Gobbler's Knob, Punxsutawney, Pennsylvania, makes his annual weather prognostication for members of the Punxsutawney Groundhog Club on February 2, 1963.

In Europe there are many sayings associated with Candlemas that are echoed in the folklore surrounding the groundhog. For example: "If Candlemas be fair and bright, Winter will have another flight; If on Candlemas day it be shower and rain, Winter is gone and will not come again." The most famous groundhog in the United States is the Punxsutawney Phil.

Canada has its own contender, Wiarton Willie, an albino groundhog, and many U.S. regions have festivals for less-celebrated rivals of Phil's. Although the record indicates that the forecasts of Phil and his fellow groundhogs are no better than chance, the custom does have some basis in meteorology. If the groundhog does see its shadow (and scrambles back into its burrow), it is because it is a sunny day. A sunny winter day at high northern latitudes often indicates a cold high-pressure system, which may persist for a few more days (though not six weeks). A cloudy winter day, which will make it more difficult for the groundhog to spot its shadow, may portend a frontal system characterized by the influx of warmer, moist air from the Gulf of Mexico. That system isn't likely to stick around longer than a few days, either.

Groundhog Day Origins
Groundhog Day, which falls on February, owes its origins in the United States to German immigrants who arrived in Pennsylvania in the 1880s. Their spring forecasting festival is based on the Christian tradition of Candlemas, but its roots extend back to ancient Rome.

The Woolly Worm

The woolly worm has long been used in certain regions of the United States to predict the severity of the coming winter based on the width and coloring of the bands that appear on the creatures. In spite of its doubtful value as a forecaster, the Woolly Worm is the focus of several popular festivals in fall.

Winter will be bad if:
Woolly worms have a heavy coat
Lots of them are seen crawling around
Their movement is unusually slow
The black band at each end is wide;
the more black than reddish brown—
the worse the winter
They are seen crawling before the first frost.

With the exception of the groundhog, the woolly worm is the best-known natural predictor of winter in the United States. The woolly worm, also known as a woolly bear, fuzzy bear, black-ended bear, or banded woolly bear, is actually the larval stage of an Isabella tiger moth (*Pyrrhactia isabella*). It passes through several larval stages before entering its pupal or winter cocoon stage. Female moths lay eggs in spring after mating; when their eggs hatch, the cycle begins once again. There are two generations of these worms each year—the first in June to July, and the second in September. It is the second brood that is reputed to be weather prophets. It grows up to 3 inches in length and can be found throughout the United States. What makes the woolly worm such a popular prognosticator is its distinctive bands: black at each end, with a reddish brown band in the

WOOLLY WORM FESTIVALS

The humble caterpillar that has inspired so much winter folklore is honored amid boisterous celebrations in several locations throughout the United States. These autumn festivals can draw up to 20,000 people. The big event of these celebrations is a woolly worm race. Attendees enter their own worms in the contest. Some of the worms are supposedly "trained," though it is unclear how you train a caterpillar. The worms are placed at the bottom of vertically mounted strings. The first worm to make it to the top is declared the winner. This is the caterpillar that is then chosen to predict the severity of the coming winter.

middle. Woolly worm caterpillars have 13 brown and black segments, which are supposed to correspond to the 13 weeks of winter.

It is the middle bands that are thought to indicate the severity of the winter. The narrower the bands, the worse the winter. Color is also important: If the terminal bands are more brown than black and the middle band more orange than brown, the winter will be mild. That is the folk belief, anyway. But researchers dismiss the idea that these worms can predict anything at all, noting that the caterpillar molts no less than six times, and that the color and size of its bands may change from molt to molt. Moreover, there are many species, some of which are solid black, some of which have no bands, and others of which have bands of varying sizes, making it more difficult to rely on them to forecast winter weather. Most species spend two weeks as caterpillars, but the woolly worm's arctic cousin takes 14 years before it becomes a moth. Some researchers have established a link between the caterpillar's color and its habitat, while others have tried to connect the number of brown hairs with the age of the creature.

When the wild geese and ducks take their leave, winter will soon be here.

Birds migrate to warmer regions for more food—but it has to be a sufficient amount of food to offset the cost in energy and lives that the flight itself entails. (Migrating birds are more prone to attack by predators.) Weather is only one, and not necessarily the most important, factor in spurring migratory behavior. Even caged birds will demonstrate increasing restlessness and show physiological changes such as increased fat as autumn wears on, underscoring the fact that migratory behavior is hardwired. That isn't to say that environmental factors don't play a part as well. Scandinavian blackbirds will migrate south to escape the harsher winters there but are perfectly content to stay where they are in winter if they live in southern Europe.

Wildfowl in the Northern Hemisphere, such as ducks, may not move a great deal if the weather is moderate. Other birds will limit their migrations. For example, the pink-footed goose will simply make a short hop to Ireland from its native habitat in England. Because most wildfowl are strong fliers, they may winter in regions with moderate winter climates; if the weather suddenly turns severe, however, they will relocate a short distance to where conditions are more favorable.

ST. BRIDGET'S DAY

One of the patron saints of Ireland, St. Bridget of Kildare was probably developed by the Catholic Church based on the earlier Celtic goddess Brigit. The goddess was celebrated by a fire festival every year called Imbolc held on February 2, which heralds the coming of spring.

The Irish have a related legend about the Old Woman of the Gloom, an evil spirit who gathered sticks on St. Bridget's Day for a fire to dry herself. If St. Bridget's Day was wet, the Old Woman stayed home; that meant that the following spring would be dry. If the feast day was dry, the Old Woman would gather sticks for a fire to keep her warm because spring would be a wet one.

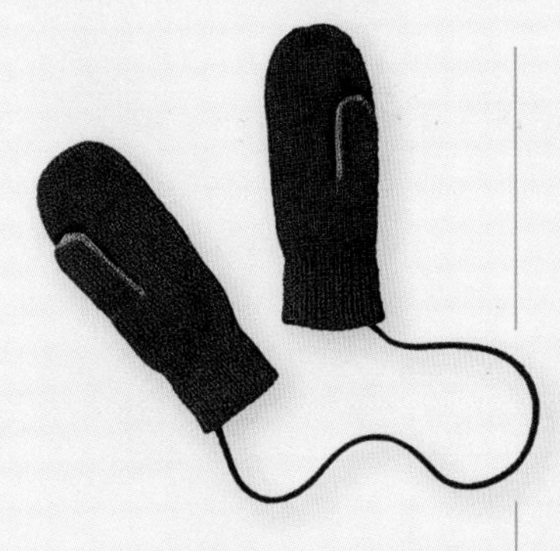

Put a coat on! You'll catch your death of cold!

In spite of the admonitions of mothers for hundreds of years, no conclusive evidence exists that cold temperature increases the chances of catching a cold. Colds (which are caused by over 200 different viruses) develop because of a weakened immune system, and temperature simply does not have any direct affect on immune systems. However, severe cold does deter people from going out of doors. With less ventilation and closer proximity to others who might be infected, people may be more prone to colds in winter months.

If the spleen of a hog be short and thick, the winter will be short.

This saying from the Pennsylvania Dutch reflects the belief that the bulges and corrugations on a hog spleen could be used to predict the severity of approaching winters. In an account of early American pioneers in Pennsylvania around 1880, one historian wrote: "Some consult the milt or spleen of the hog, that organ situate in the left hypochondrium, and which was supposed by the ancients to be the seat of anger and melancholy; and from this organ they augur the severity of the approaching winter." The practice, which has no basis in scientific fact, may go back to the ancient Etruscans, who used the entrails of animals for divination.

When the spider withdraws into its chamber, winter will follow her right away.

This saying, which comes from Germany, touches on a more universal superstition about the power of spiders. In Russia, for instance, an encounter with a spider foretold the arrival of a guest, the appearance of a hidden enemy, and good or bad news—but mainly good because a spider was said to be a harbinger of financial or other success. As a forecaster of a cold winter, spiders are reliable only to a limited extent. Like all creatures, they react to—rather than foretell—weather. Spiders try to consume as much as possible before the onset of cold weather. They lay their eggs in fall, which means that they are busy spinning webs. Because late summer can influence the early part of the winter weather, the activity of spiders might provide some indication of the winter to come. Spiders survive winter as subadults or adults. Immature spiders can survive only in warmer months.

COLD FRONTS

There are four types of fronts—cold and warm fronts, and stationary and occluded fronts. They are models, meaning that they are useful for studying and making predictions, but they do not always occur as "perfect" cold or warm fronts in nature. Cold fronts are not always "cold." The arrival of a cold front means that you are in for a spell of cooler weather accompanied by clearing skies and a sharp change in wind direction. In winter, cold fronts do indeed bring frigid air. But in summer, cold fronts often bring cooler, drier air, which can come as a relief to people who have been suffering from a bout of humid, sweltering weather. A cold front is the leading edge of an air mass of noticeably colder and drier air than the air mass it is about to enter. It can also be thought of as a transitional zone between a cooler and warmer air mass.

Cold fronts typically move at a rate of about 9 to 31 m.p.h. (15 to 50 kph). They also maintain their intensity for a longer duration than warm fronts. As a cold front passes and begins to replace the warmer air mass, it can cause a drop of temperature of about 15°F (-9.4°C) within the first hour. These fronts move faster than other fronts–about twice as fast as a warm front–which is one of the reasons they tend to produce the most violent weather, such as thunderstorms and tornadoes. They are usually preceded by cirrus clouds–thin, wispy clouds that form at high altitude, but as they get closer, they fill the skies with a large variety of clouds, especially cumuliform (puffy, cotton candylike) clouds.

Cold fronts are frequently associated with squall lines—a string of strong thunderstorms parallel to and ahead of the front. In general, precipitation occurs just behind the front. This is the result of frontal lifting, a phenomenon that occurs because colder air is denser than warmer air (meaning that its molecules are more tightly packed into the air packet) so that it tends to sink, while warmer air, being buoyant, will rise. In effect, the cold air wedges in below the warm air, closer to the ground. As the rising air cools, it expands, triggering condensation of water vapor. That in turn produces precipitation.

Cold Fronts

On weather maps a cold front's position is designated by the symbol of a blue line of triangles or spikes (pips) pointing in the direction the front is heading.

Snow

"Snow cherisheth the ground and anything sowed in it," said the thirteenth-century friar, scientist, and philosopher Roger Bacon. The saying, like so many weather proverbs about snow, recognizes the protective value that snow offers for flowers and plants during the cold winter months. Without the insulation of snow, seedlings and new growth could perish, ruining the planting season. Folklore also distinguished between different types of snow. The Finns used to believe that Snow was an ancient king with three daughters—Thin, Thick, and Snow Storm.

Stranded!
Heavy snowfalls can cause a great deal of disruption and trap motorists, but not all severe snowstorms are blizzards. Blizzards are winter storms that bring both snow and strong winds and produce low visibility. Blizzards can even form when winds pick up and blow snow that has already fallen. According to the U.S. National Weather Service, winds must attain speeds in excess of 35 m.p.h. (56 kph) and reduce visibility to less than a quarter of a mile for more than three hours to meet the definition of a blizzard.

No two snowflakes are alike.

Although it is impossible to reach a conclusion without an examination of every snowflake that has ever fallen, researchers have in fact made the effort. The first to try was the great astronomer Johannes Kepler, who wrote a book about snow in 1611. The most intensive investigation was conducted by Wilson A. Bentley, a native of Vermont, who spent 40 years photographing snow crystals, culminating in the publication of his book *Snow Crystals* in 1931. A Japanese researcher, Ukichiro Nakaya, contrived a classification system for snowflakes based on seven types of crystals: plate crystal, stellar crystal, column, needle, spatial dendrite, caped column, and irregular crystals. A Canadian scientist once declared that he'd discovered two snowflakes that were the same, but closer investigation revealed tiny variations.

If the first snow stays on the ground for three days, another snow will come top it.

According to folklore, snow drifts that remain on the ground for a prolonged period will turn into "snow breeders," which attract more snow. Although there is little evidence for this, fallen snow does have an affect on the weather. Snow cover will consume more of the sun's energy than exposed ground will; some of that energy is used to melt the snow, while some energy is deflected by the snow.

The result is that there is less energy to heat the Earth's surface and so temperatures will not rise as high as they would have if no snow had been on the ground. At night, snow readily gives off heat, which contributes to rapid cooling.

SNOW FACTS

- In a period of seven days in February 1953, a total of 187 inches of snow (474 cm) fell in the town of Thompson Pass in Alaska. Buffalo, New York, was buried by 82 inches of snow (208 cm) in just four days in December 2002.
- The highest seasonally cumulative precipitation (as snow) ever measured was on Mt. Baker in Washington State. Between 1998 and 1999 it received a staggering 1,140 inches (28.96 m) of snow.
- The only snow actually to appear on the equator is at an altitude of 15,387 feet (4,690 m) on the southern slope of Volcán Cayambe in Ecuador.
- In 2005 Algeria experienced its largest snowfall in half a century, which killed 13 people.
- Blizzards—snowstorms with winds of at least 35 m.p.h. (56 kph)—are known as buran in Russia, purga in Siberia, and boulboe in southern France.
- A snowfall of 10 inches (25 cm) can cover 3.3 ft by 100 ft (1 m by 30 m) of a sidewalk with about 1,650 pounds (750 kg) of snow.
- Snowflakes are actually aggregates of smaller snow crystals, often containing hundreds of individual crystals.
- Avalanches have been used as weapons. In World War I, opposing Austrian and Italian troops in the South Tyrol fired at snow-covered mountains to trigger avalanches, which led to the deaths of at least 10,000 soldiers on both sides within a period of 24 hours.

Snowflakes

While studies have confirmed that no two snowflakes are alike all of them have the same hexagonal structure and are formed in the same way. Snow crystals are single crystals of ice whereas snowflakes may consist of several snow crystals stuck together.

Smoke Signals

Whether smoke will rise into the air or curl when it emerges from a chimney may depend on weather conditions. When skies are clear and winds are calm, smoke will continue to rise. But under conditions of lower atmospheric pressure and cloudy weather, smoke no longer rises as quickly or as high because of the insulating effect of the clouds. In that event the curling smoke may forecast an approaching storm.

Eskimos have hundreds of words for snow.

The belief that Eskimos—more accurately known as the Inuits—have more words for snow than other peoples has persisted for many years. There are claims that the number of words is in the hundreds, but in fact this is not the case at all. In the 1991 book *The Great Eskimo Vocabulary Hoax*, the linguist Geoffrey Pullum debunked this myth. For one thing he pointed out, there isn't one Inuit language, but eight, and that the number of words for snow in all of them is about four or five. That's about the same number as in English. Moreover, the Inuit languages are polysynthetic, with many words being created from just a few roots, rendering the whole debate moot: "Where English uses separate words to make up descriptive phrases like 'early snow falling in autumn' or 'snow with a herring-scale pattern etched into it by rainfall,' Eskimo languages have an astonishing propensity for being able to express such concepts (about anything, not just snow) with a single derived word."

When dimmer stars disappear, expect rain or snow.

If you can no longer see distant stars on a night that the moon is surrounded by a halo, then you can expect a storm, because as the air grows more unstable and turbulent with the approach of a low-pressure system, it obscures the atmosphere.

If the snow coming down the chimney sounds like boots swishing through snow, it will be a deep, dry snow.

Weather can have an unusual acoustical effect. When clouds are lower, sounds travel greater distances because the cloud reflects the sound back to the ground. On the other hand, when skies are clear, sound travels upward and outward without hindrance,

THE PROBLEM OF SNOW PREDICTION

Forecasting snow is a more challenging enterprise for meteorologists than predicting rain, for a number of reasons. For one, whether precipitation is snow, rain, hail, freezing rain, sleet, or some combination depends largely on temperatures in the air and on the ground. Snow can form at one altitude, change to rain, and then back to snow before it reaches the ground. These transitional zones, as they are known, can play havoc with predictive computer models because a change of only a few degrees one way or another can change the form of precipitation. In addition, heavier snowfalls tend to fall in narrow bands—on a smaller scale than is generally monitored by weather stations. Using clouds to predict snow carries its own complications; by the time the snow descends, the cloud it came from may be long gone. Wind, too, can cause snow to blow quite a distance—up to several miles—from where it is was expected to come down. Snowflakes also land at different rates, depending on their surface area and air resistance. There are also different types of snow: While all snow is composed of hexagonal ice crystals, it can appear in blizzards or in flurries or snow showers that suddenly start and stop, changing in terms of intensity, accumulation, and coverage.

and becomes more difficult to detect. A similar effect occurs when a high-pressure front approaches; under that circumstance, the air becomes thicker and more dust particles become suspended in air, making objects appear hazier. Sound will become sharper and more focused prior to stormy weather because the dust reflects the sound, just like cloud cover does.

A storm makes its first announcement down the chimney.

In fact, chimneys can amplify sounds, so people gathered around the hearth will be able to hear a dog howling in the distance that someone standing outside the house would not detect. Many theories have been offered to account for this odd phenomenon. It is possible that sound travels better vertically than horizontally, or it might be that the heat of the chimney accelerates the circulation of air, which in turn amplifies sound. The effect also may be due to the higher location of the chimney so that it serves as kind of a megaphone, allowing it to pick up sounds that might be drowned out at ground level by the murmur of wind in the grass and the bushes. People in hot-air balloons report that they are even able to hear conversations on the ground below.

SUPERCOMPUTERS

Data streams into weather stations from around the world, generated by mountaintop observation stations, satellites, ocean buoys, weather balloons, and aircraft. In the United States alone, there are more than 200 million weather observations every day. Computers make it possible for weather services to produce information, and use complicated regional and global models based on previous weather patterns to forecast possible weather conditions in the future.

Supercomputers can perform 7.3 trillion calculations per second, or 7.3 tetraflops. But computer developers have set their sights on speeds measured in the petaflops—the equivalent of processing data contained in a stack of paper 62,140 miles (100,000 km) high—every second!

Red snow gives a bloody liquor when it is squeezed.

The phenomenon of red snow is rare, but it has occurred frequently enough to warrant the attention of historians. The philosopher Aristotle mentioned red snow in the third century B.C. The Roman historian Pliny believed that red snow was the equivalent of rust. The English historian Thomas Short, when reporting on a red snowfall in Genoa, Italy, made the observation that when the unusual snow was squeezed, it produced bloody liquor. A 6-foot (1.8-m) "bloody" snowfall reportedly occurred in the Alps in 1755. In 1895 the town of Alma, Colorado, was visited by a pink snowstorm. An observer said that residents' clothes were covered with material similar in consistency to mud. Red snow is also called watermelon snow because its odor is faintly reminiscent of the fruit. It is found during the summer in the Sierra Nevada in California at altitudes of 10,000 to 12,000 feet (3,000–3,600 m), high enough for temperatures to remain cold year round so that the snow lingers. Stepping on this snow will make it redder and leave reddish stains on the soles of shoes or boots. The cause of the coloration is a species of algae that can flourish in freezing water and contain a bright red carotenoid pigment. Its Latin name is *Chlamydomonas nivalis* (nivalis refers to snow).

There have been episodes of red snow that appear to have a different cause, related to the catastrophic dust storms that struck the Midwestern United States during the 1930s. In 2006 a newspaper in Moscow carried a report of creamy pink snow falling on the northern regions of Russia's Maritime territory, which was attributed to a powerful cyclone that had gathered up sand from neighboring Mongolia. There have also been instances of yellow snow. In one case in March 1879, its color was traced to the presence of pollen from pine trees. In another case a yellow snowfall in Michigan in 1902 was ascribed to a mixture of loess, a clay-based soil in Wisconsin and Iowa, which had been carried farther east by a windstorm.

When the snow falls dry, it means to lie; but flakes light and soft bring rain oft.

Dry snow is the powdery kind: It is snow that cannot be compacted, ruling it out as a source of snowballs. Dry snow will remain dry so long as the temperature is below freezing. Wet snow, on the other hand, is snow that contains a good deal of water; the more water it has, the wetter the snow. Dry snow is more likely to be responsible for avalanches because it isn't amenable to compacting. Dry powdery snow can become moist due to a thaw or rainfall. Whether dry snow stays put as the saying contends, or undergoes change while it lies on the ground, depends on a great many factors including changes in energy, wind, moisture, water vapor, pressure, and its constituency. Snow on the ground is a mixture of ice, liquid water, and water vapor, and that mixture can vary enormously. It is a common misperception that snow is dynamic only when it is falling; it remains dynamic when it is lying on the ground. For example, a succession of warm, sunny days may melt the snow surface, producing a high-density, well-bonded layer, while just below it the layer is colder and not well bonded at all.

Collapsing Power Lines
This massive ice storm in Canada in 1998 was the country's most expensive natural disaster. It caused about 1,000 electricity poles to collapse under the weight of the ice.

FREEZING RAIN AND ICE STORMS

Freezing rain is very dangerous. It occurs when the air is above freezing, the ambient temperature is close to freezing, and the ground temperature is below freezing. A freezing rainstorm, which produces a treacherous glaze on roads, is somewhat different from an ice storm, which is the result of rain that has been supercooled and freezes on impact with cold surfaces. It does form as ice in the atmosphere. An ice storm will also drape a layer of ice on plants, crops, structures, and telephone and power lines, causing a good deal of damage in the process.

A severe ice storm in 1998 in eastern Canada and upstate New York caused 3 million people to lose power, some for several weeks. The ice was 3–4 inches (7–11cm) thick in places. Ice storms are less climatically discriminating than snowstorms because they don't require very cold temperatures, which is why they can occur in warmer regions such as the southern United States. They have even been known to ruin citrus crops in Florida.

If the snow remains on the trees in November, it will bring out but few buds in the spring.

This saying, which originates in Germany, has a kernel of truth in it: Snow does have an impact on the budding and flowering of trees in the spring. European researchers—dendroclimatologists who are specialists in the study of tree rings—undertook a study of Siberian trees to investigate the slowing growth of trees since the 1960s across subarctic regions from Alaska and Canada to Scandinavia and Siberia. Tree rings do provide information on seasonal and annual changes in temperature and precipitation as well as long-term changes. The researchers' study of trees in Siberia was designed to find out why they were not growing as quickly as expected, even though average temperatures in spring and summer had been higher than usual during that time. What they discovered surprised them: A greater accumulation of snow was keeping the ground frozen longer, and this was retarding the spring greening. The Siberian trees were not growing as fast as possible because the ground in which they were rooted was not thoroughly thawed out during the summer. Snow insulates trees in winter, but it also insulates a vast amount of solid ice for long stretches of time. In Siberia, the spring thaw does not arrive until June, leaving less time for the sun's energy to reach the icy soil and consequently trees have less growing time.

WHY ARE SNOWFLAKES SYMMETRICAL?

The symmetrical form of snowflakes is due to the way that their water molecules are arranged in their solid or frozen state. The process of snowflake formation is known as crystallization. Water molecules form weak hydrogen bonds to produce the hexagonal shape that characterizes all snowflakes. The arrangement is designed to maximize attractive forces and minimize repulsive ones. The process is similar to the way tiles fit into a preselected pattern on a floor. The tiles have to be placed in a certain order if the pattern is to be retained. That is what water molecules do: They take their places in a prearranged pattern. This is not to say that snowflakes are alike—just the opposite. Snowflakes may have the same form, but the environment affects them in different ways: Crystals can be molded by atmospheric conditions, humidity, and temperature differences.

The number of days the last snow remains on the ground indicates the number of snowstorms that will occur the following winter.

This saying is an example of seasonality that refers to nonbiological events—when leaves fall or frost appears, for example. It is related to phenology, which is based on the idea that periodic plant and animal life cycle events are influenced by environmental changes. The opening of flowers, the migration and return of birds, the hatching of insects, have all been applied to forecasting future weather conditions. Advocates of phenology contend that by scrupulously recording the behavior of animal, insect, and botanical life, they are capable of monitoring and measuring the impact of local and global changes in weather and climate on the earth's biosphere. There is, however, no evidence that the length of time that snow remains on the ground in a particular area has any predictive value about how many snowstorms to expect the following winter.

If snowflakes increase in size, a thaw will follow.

No one really knows how large snowflakes can grow. Snowflakes, which are agglomerates of many snow crystals, are less than half an inch (12 mm) across. Most heavy snowfalls occur in conditions with relatively warm air temperatures near the ground—typically 15°F (-9.4°C) or warmer—since air can hold more water vapor at warmer temperatures. However, when temperatures are near freezing, winds are light, and atmospheric conditions are unstable, much larger snowstorms can sometimes produce irregular flakes close to 2 inches (5 cm) across. So larger flakes are more likely to be produced when colder conditions prevail. There is no reason to think that they predict a thaw.

Corn is comfortable under the snow as an old man is under his fur coat.

This Russian saying means that snow acts as a thermal insulator, conserving the heat of the earth and protecting crops from subfreezing weather. There is a growing apprehension among many international scientists that global snow cover is being diminished by climate change, although the data in many parts of the world remain inconclusive. A study of alpine environments indicated that if snow cover were reduced through climatic warming, plants that are normally protected in winter would become exposed to greater extremes of temperature and solar radiation.

Winter snowfalls occur frequently in Russia, but they are rarely very heavy. Strong winds that can accompany snowstorms, producing blizzards, can clear away more snow than they leave behind. Rivers in much of Russia, however, are frozen from 70 days a year in the west of the country to as much as 250 days in northern Siberia. But even the famously harsh Russian winter, which contributed to the defeat of Napoleon's army in 1812, is beginning to change in response to global warming. In 2006 the country experienced its warmest November and December since records have been kept. Mushrooms sprouted out of season, and bears were unable to hibernate.

Nine months of snow followed by three months of bad tobogganing.

This old Yankee saying, while obviously an exaggeration, underscores the problematic weather in New England, a region of the United States comprising Maine, New Hampshire, Vermont, Massachusetts, Rhode Island, and Connecticut. The more uplifting saying New Englanders like to cite is, "If you don't like the weather, wait a while." But winters in that area do tend to be especially snowy.

Tobogganing, or sledding, is a winter sport and pastime whose origin can be traced back to Native Americans. For the Innuit and Cree of northern Canada, it is still a traditional form of transport. A toboggan is a flat-bottomed vehicle of hard wood with a waxed bottom that is steered down a snowy incline by a rider shifting his body weight. Tobogganing has been modified for the Olympic sport of bobsledding, which takes place on tracks of ice.

WINTER FESTIVALS

Every winter in different parts of the world thousands of visitors are drawn to festivals that feature extravagant snow sculptures.The city of Sapporo, on Japan's northernmost island, Hokkaido, hosts a festival that offers a showcase of ice maidens, gods, demons, palaces, and pyramids. Up to 2 million people attend every year.

Ice and snow sculptures along with canoe and dogsled races are featured at the annual Quebec Winter Carnival, the world's largest winter festival and the third largest carnival in the world after Rio's Carnival and the Mardi Gras in New Orleans. Harbin in northern China and Perm, Russia, also host major festivals.

Snow squeaks when it is very cold.

Cold temperatures can affect sound just as they can the formation of snow on the ground. The louder the squeak, the lower the temperature is likely to be. When the temperature of the air and snow are only slightly below freezing, the pressure of the foot compresses and partially melts the snow crystals. Because it is lubricated by a thin film of water, the snow can flow, and little sound can be heard. If temperatures are sufficiently below freezing, the foot pressure is not enough to cause any melting. According to one theory, the pressure causes the packed ice crystals to crash into one another. That smashing produces the squeak. Another theory holds that the pressure causes the snowflakes to expel air that is trapped between the crystals, and the air makes the noise.

Body Temperature and Cold
Hypothermia is defined as a drop in body temperature. In response, the body will try to generate more warmth, which results in shivering. In more extreme cases, the body will begin to curtail blood flow to the extremities to preserve heat.

HYPOTHERMIA AND FROSTBITE

In ancient Japan it was believed that a spirit called Snow Woman would put tired travelers to sleep after they had wandered too long in frigid weather, spelling their doom. Extreme cold can result in a condition called hypothermia if the victim is inadequately protected. The degree of exposure is directly tied to the severity of the condition. Hypothermia occurs when the core body temperature drops below normal. Warning signs include uncontrollable shivering, disorientation, incoherent or slurred speech, and cold pale skin. Left untreated, it will cause drowsiness, extreme confusion, and slowed breathing. If the condition persists, unconsciousness and then death will follow. Treating hypothermia involves warming the trunk (the body) before the extremities.

Frostbite is the damage produced by extreme cold to exposed skin. There are three stages of frostbite. The first, known as frostnip, will cause numbness in the exposed areas (typically fingers, toes, earlobes, and the tip of the nose) and give them a white appearance. Continued exposure will cause the skin to feel hard and frozen, with the possibility of blistering, a stage known as superficial frostbite. In the third and most serious stage, the skin will turn blotchy and blue and affect underlying tissues. Anyone suffering from frostbite is urged not to rub the frostbitten areas because that will only exacerbate damage.

Wind Monitor

Wind monitors are sensors that measure wind speed and direction. The wind sensor is a four-blade helicoid propeller whose rotation produces a voltage signal. The signal's frequency is directly proportional to wind speed.

It is too cold to snow.

The reason that snow might not fall on a very cold day has nothing to do with the temperature; rather it has to do with the dynamic stability of the atmosphere. Only when truly frigid arctic temperatures prevail—around -40°F (-40°C) and below—is the moisture capacity of the air so low that very little snow can occur. Otherwise, even when the surface temperatures are very cold, significant snowfall can occur.

Snow cannot fall unless certain conditions are met. Generally, the temperature has to be below freezing, the air has to be saturated with moisture, and enough air must be rising from the ground to allow moisture-laden clouds to develop. This saying refers to conditions when the third criterion is not met—there is not enough lifting. To lift air, some force is required. Wind carrying cold air can supply that force, driving a wedge to lift relatively warmer air at the surface. Wind can also bring more moisture into a region, providing an additional ingredient for snowfall. However, when conditions are stable—winds are light or nonexistent—the force is absent. The air on the ground stays there. Meanwhile, as temperatures drop, the air is capable of holding less and less water vapor, making snowfall less likely, although it always contains *some* moisture (at least until it reaches absolute zero).

Plants show that winter will be rough if the crop of holly and dogwood berries is heavy.

Both holly and dogwood are associated with winter. In most myths, the holly tree is paired with the oak tree. The dogwood seems to have slipped into this saying

PROFESSIONAL WEATHER STATIONS

A weather station is an observational post that monitors atmospheric conditions and collects meteorological data using an array of instruments and sensors. There are two types of weather stations: manual and automatic. Manual weather stations must be staffed. Automatic weather stations (AWS) are generally used when there is no one available to staff them, or when the station is too remote to be easily accessible. These stations may report weather conditions in real time via global communications or else store data for recovery later. Most AWS have the same basic equipment found in manual weather stations, although they have certain limitations—they cannot report on cloud types or the amount of moisture in a cloud, since this requires human observation.

because its berries do not stay on the tree as long as other fruits and berries, which incidentally deprives bluebirds of a favorite source of food. By midwinter, dogwood trees are usually bare. The dogwood berries ripen around mid-October to early November. An unusual number of dark berries on the trees doesn't signify a harder winter than normal; it means that the weather has been warmer for a longer time in fall, leading to more growth, but also to more infestation from worms.

In the Christian tradition, holly became associated with Jesus not only because of his birth around the time of the winter solstice, but because his crown of thorns was made out of holly leaves. There are some etymologists who maintain that holly is a corruption of the word "holy," although this is disputed. One thing that is not in any dispute is that holly comes in many varieties—there are approximately 400 species—and is one of the few plants that can flourish in a variety of climates. For instance, holly can grow successfully in every one of the 50 U.S. states, including in the widely different climates of Hawaii and Alaska. Holly species are found on all continents, with the exception of Australia and Antarctica. These plants also show a great diversity in their shape and size, too, ranging from a dwarf shrub 6 inches (15 cm) high to a giant holly tree of 70 feet (21 m) tall.

Holly
The holly has long been associated with winter and Christian and Druidic traditions.

ICE FOG AND FREEZING FOG

Liquid needs a surface to freeze on. That is why water vapor suspended in the air (a cloud) doesn't always crystallize even when temperatures are below freezing—as low as 14°F (-10°C).

Liquid droplets will freeze even when no surface is available if the temperature drops low enough to below -40°F (-40°C). Those drops that remain liquid below freezing are called supercooled. Freezing fog is composed of supercooled droplets.

Ice fog is composed of tiny ice crystals, not liquid droplets, which means that temperatures are too low for the droplets to be supercooled, conditions that are typically found in Arctic or polar air.

If October bring heavy frosts and wind, then will January and February be mild.

After enduring long periods of snow and ice, people begin to long for thaws. Thaws can occur at any time during winter; and "January thaws" in the cold, snowy portions of the United States are not uncommon, although they cannot be depended on. These winter thaws are similar to Indian summers (which occur after the first killing frost) in that they are unpredictable. Meteorologically speaking, these warm intervals may be related. Both weather patterns are often produced by a combination of highs and lows that results in an influx of tropical maritime air with above-average temperatures, hazy skies, and gentle southerly winds. In Iran the snowy season is also interspersed with thaws (described as "snow eaters" if they are rainy), which periodically bring about melting of the normally thin snow cover. In Scottish lore there are three cold winds, one of which is known as "wind before thaws."

Watch out when the gray mare's tail begins to grow.

Thaws are not altogether welcome, because melting snow can bring danger. In Scotland, people inhabiting mountainous terrain are put on guard when they see waters streaming down from the heights that contain white streaks—the "gray mare's tail"—a sign of melting snow. A sudden surge of water that submerges previously dry land can suddenly alter their travel plans and leave them stranded.

The Cuidh-Crom begins to break when a thaw is on its way.

The Cairngorm in the Scottish Highlands is covered by what is described as a great snow wreath called the Cuidh-Crom—"the bent or crooked wreath." During a thaw the snow begins to crack until the whole structure crumbles. If the Cuidh-Crom does not break apart in May, and if the whole wreath has not melted away by the middle or end of June, then the locals call it a late season.

The Cross begins to bend when the snow is over.

The saying above refers to thaws in the high Andes. The Cross is the constellation of that name that appears in the Southern Hemisphere in spring.

TYPES OF SNOWFALL

Not all snowstorms are alike. For that matter, not all snowstorms are really storms. Flurries are defined as snow that suddenly stops and starts and can abruptly change in intensity; accumulation and coverage are limited. A snow squall is basically flurries along with strong winds. Snow can blow and drift—and there are definitions for each—but they aren't storms.

Blizzards are defintely storms. They are defined as severe snowstorms lasting three or more hours that bring with them low temperatures, strong winds of at least 35 m.p.h. (56 kph), and poor visibility because of blowing snow. Some blizzards attain wind speeds over 50 m.p.h. (80 kph). Blizzards are formed by a clash of warm and cold fronts. Some regions are more prone to blizzards than others, even in places that are accustomed to cold, snowy winters. South Dakota achieved a dubious record of 265 blizzards in a 20-year period (1978 to 1999), while the United States as a whole experiences an average of 12 blizzards every winter. In addition to paralyzing communities for days at a time and exacting considerable economic losses, blizzards have claimed many

If the foehn did not interfere, neither God nor his sunshine would ever be able to melt the winter snows.

The foehn is a local wind that blows in mountainous regions, most commonly in the Alps. It brings respite from the winter cold and is very effective at melting snow, a fact reflected in many sayings from the Alpine region. These winds move upslope, expanding and cooling. In the process, precipitation is released, removing moisture from the air. Then the dehydrated air passes over the crest of the mountain, and in its movement down the other side of the mountain its temperature increases under greater atmospheric pressure, creating strong, gusty, warm, and dry winds.

April snow is as good as a lamb's manure.

Even as spring begins to assert itself, there is always the prospect—at least in the United States and Canada—that a snowstorm may bring a nasty surprise, an event that generally evokes feelings of disgust and frustration from people who had thought they had gotten winter out of the way, as this saying makes very clear.

lives: In March 1993, a two-day blizzard killed 270 people on the eastern seaboard of the United States. The most famous blizzards of all struck the country in 1888, when one storm pounded the Northeast and another the Midwest. The latter became known as the Schoolchildren's Blizzard because of the 235 people who died as a result, most were children.

There is also a form of snowstorm that is peculiar to the southern and eastern shores of the Great Lakes of North America, Lake Baikal in Siberia, and elsewhere, called lake-effect snow. The phenomenon results when cold arctic air moves over warmer water and gathers water vapor from the lakes. The water vapor condenses, forming clouds and producing snow that falls on the shore downwind.

Surely some of the strangest snowfalls have involved worms and insects—an anomaly dubbed "bug snow." Bug snows have been reported in Russia and Eastern Europe even into the twentieth century. In 1922 a snowfall containing spiders, caterpillars, and large ants descended on the Swiss Alps. Observers at the time speculated that the insects had been "blown in on a wind from a warmer climate."

FROST AND HAIL

FATHER FROST
In Russian mythology, Father Frost was a blacksmith who forged great chains of ice to bind water to the earth each winter. As in many fairy tales, the villain is a stepmother. The stepmother has two daughters, one her own and one her unloved stepdaughter.

One day she persuades her husband to leave his daughter out in the cold to freeze to death. But Father Frost finds her, and upon discovering that she has a good heart, he protects her, then sends her home with precious jewels and beautiful garments. When the stepmother realizes that her plot has failed, she leaves her own daughter out in the fields in hope that Father Frost will reward her as well. But she is rude to Father Frost, who becomes angry and freezes the girl to death.

Frost abounds in the myths and folklore of many cultures, possibly even more so than snow. In Finland and northern Russia, frost takes the form of a pair of deities—Frost Woman and Frost Man—who control blizzards. In Norwegian mythology, Jokul (or Frosti), son of Kari, the god of the winds, would blow frost on the earth when he was angry. Jack Frost probably owes his origin to the Norse myth. Only after reaching British shores did Jokul Frosti become Jack Frost, an elflike creature who gives color to leaves and traces intriguing patterns on windowpanes.

If in January the ice and snow crunches,
at harvest time there'll be grain and clover in bunches.

This saying reaffirms the belief that so long as the winter is behaving in predictable ways, a good planting season and bountiful harvest are assured. It is unseasonable weather that causes alarm. This saying, like many other winter weather lore, is based on the observation that snow acts as an insulator, protecting seeds and new growth from frigid temperatures. If the snow cover is inadequate, or if the weather turns unseasonably warm and melts the snow, plants and crops are vulnerable to a killing frost.

Frost or dew in the morning light
shows no rain before the night.

Frost is an accumulation of ice crystals on cold surfaces that forms when the air cools at night. Vapor condenses out of the air, coating surfaces with water if the temperature is just above freezing (32°F, or 0°C)—the dew point. It's for this reason that frost has been called a "first cousin" of dew. If it drops below freezing, though, the condensed water vapor freezes and produces frost. (There are certain conditions when it can form at slightly above freezing.) Since cold air is denser than warm air, frost tends to gather in low areas or hollows. In order to form, both frost and dew require cooling temperatures and clear, calm weather at night. Nights with such conditions are generally followed by pleasant days. So the saying is correct, with the qualification that there is always a possibility of a storm system arriving later on that will turn a fair day foul.

The first killing frost comes precisely 90 days after the first katydids begin to sing.

Katydids usually begin to sing in midafternoon by late August in the Northern Hemisphere. They are related to crickets and grasshoppers, and in fact resemble a green grasshopper with what one writer describes as "a built-in fiddle that plays only a three-note tune." It actually produces the sound of that "fiddle" by rubbing its wings together. Although the song of the katydids heralds autumn, they are unreliable predictors of the first harsh frost, generally falling short of the mark.

If hoar frost comes on a morning twain, the third day surely shall have rain.

This saying reflects the evanescent nature of hoar frost that tends to form on branches, wires, poles, and other objects. Hoar frost develops in a manner that is similar, though not identical, to the way that dew does. Both dew and hoar frost are deposited on objects where the air is incapable of retaining water vapor any longer.

Frost consists of crystalline structures formed from water vapor that has evaporated and condensed. It can remain as long as conditions are favorable, but if the air or crystals in the air around the frost become warmed, it will quickly evaporate. That happens frequently in the late winter because the sun is higher in the sky and its rays are stronger. So while the disappearance of hoar frost on a third day does not necessarily mean that rain is on its way, it may often indicate a trend of warmer weather; so if precipitation does come, it is more likely to be rain than snow.

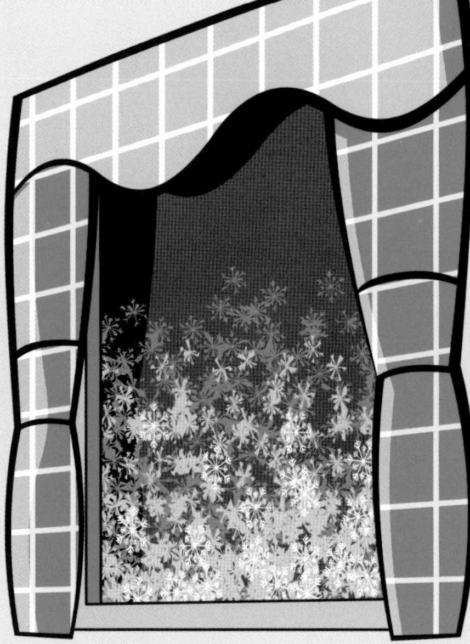

THE SEVEN SISTERS

In the myths of native Australians, the origin of frost is attributed to seven beautiful—and dangerous—sisters called the Maya, who dwelled in the sky and hurled down icicles.

Frost developed from these icicles. In spite of their allure, the sisters were unable to find love on Earth; they made their home in the heavens, forming the seven stars of the Pleiades constellation.

They would roam Earth with their hair flying behind them like storm clouds. Men would only have to catch sight of them to fall in love, but the hearts of the sisters were as cold as the ice that draped their bodies. If they were angry, they would journey to the west, where winter was to be found, and send frost and severe weather to punish those who had offended them.

So many mists in March, so many frosts in May.

The association between mist (or fog) and frost has long been known. In Japanese mythology, for example, Frost Man's disreputable brother is Mist Man. To some degree this association is based on science. In March, fog radiates heat into space during the night before being dissipated by the sun early in the morning. This effect is seen on calm, clear nights under high-pressure systems. These same conditions can also prevail in May, promoting the formation of

RIME FROST: WHITE DEATH

There are two types of frost—rime and hoar. Rime frost occurs when the rate of frost formation is rapid, typically when there's high water content in the air or when clouds pass over subfreezing surfaces—coastal mountains in winter, for example. Accumulations of rime frost can be several feet (1+ m) thick and develop in the direction of the prevailing wind. The Shoshone Indians of the Rocky Mountains called rime frost "pogonip," meaning "white death." Sometimes pogonip takes the form of tiny ice crystals suspended near the ground as mist or fog, when warm air from a valley rises up and meets cooler air from the mountaintops. The resulting ice fog can come in rapidly and last for days at a time. Pogonips can be quite dangerous, too, which probably accounts for the Shoshone name. One witness, reporting on an incident of what he called "ice smoke" in Greenland in the 1880s, said that it could produce blisters on exposed skin and "was very pernicious to health."

There is another form of frost called hoar frost, which forms when water vapor accumulates slowly on a surface, but only when winds are light. Starting as a small seed, it develops into a feather, form, or floral pattern. Hoar frost can be expected on cold, clear nights and is the most common frost in nonmountainous areas. Hoar frost has a crystalline pattern, in contrast to the more chaotic pattern seen in rime frost.

ground frost if the temperatures are cold enough. So in effect, conditions must be the same in both months for a connection to be established. That is why a clear moon is a precondition of frost. Inclement weather is decidedly unfavorable for its formation. However, there is no way to ensure that the number of mists in one month will be equal to the number of frosts in another.

There is some reason to think that this saying originated in the English county of Wiltshire, because it is located deep enough in the countryside where ground fog tends to occur. Researchers have actually compared the number of days of fog in March and days of ground frost in May in two villages in the county—Lyneham and Boscombe Down—over a 25-year period, a rare instance in which weather lore can be checked against the record. The results were not very encouraging if you believe in the lore: The number of fog and frost days in a year was equal in Lyneham only once in a quarter of a century, and in Boscombe Down, only twice.

A high wind drives away the frost.

This saying is not precise, but it does reflect a reality. Wind cannot dispel frost; however, wind can curb the formation of frost. To develop, frost needs cooling temperatures, whereas wind can bring warm air, slowing the cooling process. Moreover, wind can stir up the atmosphere, making it more difficult for radiation to escape into space.

The Lord rained hail upon the land of Egypt. So there was hail, and fire mingled with the hail, very grievous; and the fire ran along upon the ground.

While other features of weather have positive associations or evoke ambivalence, depending on when they occur, almost nothing favorable has been said about hail. Hail usually occurs as a byproduct of violent thunderstorms. In the Judeo-Christian tradition, hail is the seventh of the 10 plagues that God visited upon ancient Egypt in an ultimately successful effort to persuade the pharaoh to free the Hebrew slaves. However, it was the only plague in which God gave advance warning: "Send therefore now and gather thy cattle, and all thou hast in the field, for upon every man and beast which shall be found in the fields, and shall not be brought home, the hail shall come down upon them, and they shall die."

HAIL FORMATION

Hail occurs in association with thunderstorms, particularly supercell thunderstorms—systems that can contain three or four thunderstorms acting in concert. Water droplets in the upper atmosphere become supercooled, and when they collide with dust or dirt particles, the droplets will freeze to them, forming seeds of hailstones.

Thunderstorms are made up of rapid updrafts and downdrafts of air, sending embryonic hailstones up and down, setting the stage for more collisions. As more supercooled water clings to them, they continue to grow until gravity exerts enough force to overcome air resistance, and then they fall. Whether hailstones retain their form as a hailstone will depend on the temperatures in the atmosphere through which they fall.

FROST AND ICE: WHAT'S THE DIFFERENCE?

Frost and ice are both formed from the crystallization of liquid water, usually when temperatures fall below freezing (32°F, 0°C). Frost consists of water molecules freezing to a foreign solid matter. It does not pass through the transitional water phase that normally occurs between a gas and a solid. Frost typically leaves a white deposit on vegetation, rooftops, and other exposed surfaces.

Ice is the solid form of water. Ice can also be formed at temperatures above freezing in pressurized environments, and water will remain a liquid or gas until -22°F (-30°C) at lower pressures. However, ice formed at high pressure has a different crystal structure and density than ordinary ice.

Early frosts are generally followed by a long, hard winter.

While early frosts do not invariably portend a harsh winter, they can be disastrous nonetheless. First frosts usually occur under a clear sky and calm winds. These are known as radiation frosts. The other type of frost, advective frost, occurs when a cold front sweeps into an area with cold and blustery winds. This kind of weather is more likely to occur later in autumn. Frosts damage plants by freezing the water in their cells, creating crystals that enlarge and expand and rupture the cells. The crystals melt in the sun, but there is no undoing the havoc that they have caused. The longer the frost remains, the more devastation it can do. Some researchers believe that prolonged exposure of several hours with an air temperature of 28°F (-2°C) should be considered a killing frost. It can take up to 24 to 48 hours after the frost has passed for plants to show signs of damage. Plants differ in their tolerance, with older crops showing more resilience than younger ones.

The sky turns green in a storm when there is hail.

There is a kernel of truth to this saying when hail is generated by thunderstorms. Why the sky turns green is not completely understood. According to one theory, clouds will turn greenish if they contain a very high amount of liquid water and possibly hail, and if the thunderstorm forms around sunset. Liquid water, like ice, is slightly blue (meaning that it is a weak absorber of red light), although the blue cannot be seen unless there is a great deal of water. At sunset the sky often turns red or orange because atmospheric particles (including dust and pollutants) scatter the blue light more than the red. When the red or orange light strikes the blue light of the liquid water in the clouds, the theory goes, it results in the eerie greenish hue that frequently foretells a storm, though not necessarily hail.

Hail brings frost in its tail.

This saying seems to conflate two different processes: Hail is produced in the violent updrafts and downdrafts of thunderstorms in summer, while frost accumulates in colder weather, when water condenses out of air and crystallizes on the ground or objects. Thunderstorms, however, may usher in a cold front that under certain conditions could briefly produce frost. Frost develops at night and, in a season conducive to thunderstorm formation, would quickly evaporate

in the morning. In the winter, storms may move ahead of a warm front. Under those circumstances, warm air flows over cold air near the ground. The result is a layer of warm air with temperatures above 32°F (0°C) sandwiched between a layer of colder air near the ground and a layer of colder air higher up.

Some precipitation that began as snow in the higher reaches of the atmosphere may turn into rain. It may remain as rain—assuming temperatures are sufficiently warm all the way to the ground—or else it may hit a colder area, turn into freezing rain or sleet. However, thunderstorms can occur along with snowstorms, and in rare instances produce a bizarre mixture of downpours, thunder, lightning, snow, and hail.

Hailstone Anatomy

The layers of a hailstone tend to alternate between opaque and clear ice. Opaque ice is formed from supercooled liquid droplets that freeze on impact, which accounts for the milky appearance.

Clear ice is also formed from supercooled liquid droplets, but because the process of freezing is slower, the air in the droplets has time to escape, so the ice is clear.

ICE CHUNKS

Mysterious ice chunks have sometimes tumbled out of the blue, and scientists are still trying to puzzle out why. In recent years California experienced such a bizarre phenomenon—twice in one week—when a chunk of ice described as the size of a microwave oven dropped down in the San Bernardino County town of Loma Linda and broke into opaque, white chunks. Witnesses said that it sounded like the detonation of an artillery shell when it landed. But they observed no plane in the vicinity, which added to the mystery.

A week earlier another ice ball had plunged down. There was no plane in sight then, either. An airplane might be a convenient explanation, but scientists agreed that even if one had been flying over the two affected areas, such a phenomenon would be unlikely. Ice that leaks from a plane is typically bluish, not white. Similar ice falls have been recorded in Spain and elsewhere. The Planetary Geology Laboratory in Madrid has collected reports of 40 cases around the world since 1999 of falling ice, or "megacryometeors" as they're called. One researcher believes that the ice was formed in the upper atmosphere by a process similar to the way that hailstones are made. The only difference is that there was no thunderstorm in the cases being studied.

It is possible that these ice chunks develop because of climate change. According to this theory, the troposphere has been expanding upward so that it reaches farther into the space than before. That might mean that its uppermost part acts like a freezer, churning out chunks of ice, some of which drop to earth.

3

SUNSHINE, HEAT WAVES, AND DROUGHTS

Nature heralds warm weather in countless ways, although meteorologists and folklorists might differ about which are the most important signs to pay attention to. Honking geese, high-flying swallows, and flowering plants all enjoyed a reputation as harbingers of spring in the past, but farmers were always on guard for too much of a good thing: Pleasant days in late spring could turn into dry, hot, humid days of summer, conditions that could spoil the harvest and cause drought.

Flock of Crows

Migrating birds—including crows—fly higher when weather is fair and winds are calm because their altitude limit is higher.

FAIR WEATHER

Marvelous weather, clear skies, light breezes, balmy temperatures… what could be better? But beautiful days tend to invite suspicion. "The morning sun never lasts a day," one saying dourly insists. "A gaudy morning bodes a wet afternoon," says another. Elusive as they are rare, days of dependably fair weather are coveted prizes, especially for farmers and sailors who studied birds and insects or scanned the skies for signs that tomorrow would be better than today, always on guard against complacency: The loveliest days could often prove the most deceptive, after all.

One crow flying alone is a sign of foul weather; but if crows fly in pairs, expect fine weather.

This saying applies not only to crows, but also to many flocks of birds whose migrations tend to be daily and take place over relatively short distances. Therefore, a crow flying on its own is not a good sign.

GLOBAL TEMPERATURES

The highest temperature ever recorded was 134.0°F (56.7°C) on July 10, 1913, in Death Valley, California, and the lowest in Verkhoyansk, Siberia, which plunged to -93.6°F (-69.8°C) on February 7, 1892. Temperatures have been rising in recent years, which most researchers believe is due to climate change. In the twentieth century, Earth's average near-surface atmospheric temperature rose about 1.1°F, ± 0.4°F (0.6°C, ± 0.2°C). This may not sound very much, but these changes may lead to more frequent—and more intense—weather, such as floods, droughts, and heat waves, as well as hurricanes and tornadoes.

Some regions around the world may experience higher agricultural yields—the northeastern United States, for instance—but others will produce far less. The retreat of glacial ice sheets, which is already happening in the Arctic and in the Alps, may result in reduced water supply in the summers, species extinctions, and increasing incidence of disease, as animal or insect carriers are able to thrive at more northerly latitudes than before. The warming trend is expected to continue, but because of climate change, some parts of the world—Scandinavia, for example—may actually experience lower temperatures on average.

Frequent heavy dew keeps the heavens blue.

Heavy dews occur on clear nights with settled conditions, which usually indicates that the following day will be clear and dry. The dew forms as the ground cools and radiates its heat back to the sky. As the surface temperature drops to the dew point—that is, the temperature to which the air must be cooled to reach saturation—the water vapor in the air is deposited on the ground as dew. At sunrise the dew evaporates as the heat grows stronger.

No weather is ill, if the wind be still.

Calm conditions, especially with clear skies, indicate the dominance of a high-pressure system. It is true that winds are stronger near the frontal boundaries of storms. Where they are absent or weak, precipitation and cloud formation are much less likely. But the saying is hardly applicable to all seasons or in all weather conditions. The phrase "the calm before the storm" is well known. Thunderstorms frequently develop in environments where winds are low. A thunderstorm cell off to the west often sucks up the westerly winds, resulting in a calm; but the calm is illusory, a temporary by-product of a developing storm. Calm conditions can also occur on very cold days with clear skies; people shivering with cold might be disinclined to think that a still wind bodes no ill.

When ladybugs swarm, expect a day that's warm.

This saying may confuse cause and effect: The insect behavior is governed by weather conditions to a large extent, but seldom actually predicts it. Nonetheless, the saying about the ladybugs may have some credibility. Ladybugs store heat in their shells. So on days that are heating up, they will start flying to dissipate the heat. In general, however, the insect swarms in summer are governed more by reproductive behavior than by current weather conditions.

SUNSCREENS

Sunscreens are essential to shield the skin from exposure to harmful ultraviolet rays that can cause sunburn and lead to tissue damage. It is important to check the label, since not all sunscreens are alike.

Of the three types of ultraviolet radiation—ultraviolet A, B, and C—UVB is the most harmful and the one most likely to cause sunburn and lead to skin cancer. A sunscreen with a high sun-protection factor (SPF 15+) is necessary to protect you from UVB.

Doctors caution that a sunscreen should not give anyone a false sense of security. Would-be sunbathers are advised to stay in the shade as much as possible and wear hats, T-shirts, and sunglasses.

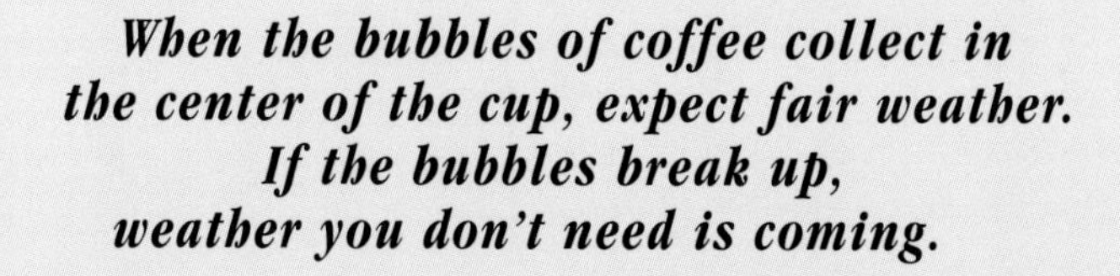

When the bubbles of coffee collect in the center of the cup, expect fair weather. If the bubbles break up, weather you don't need is coming.

There is some justification for thinking that you can predict the weather by staring into your cup of coffee in the morning. High air pressure associated with fair weather will cause the surface of the coffee to become slightly concave so that the bubbles will gravitate to the middle of the cup. By contrast, low air pressure during foul weather will make the surface of the coffee slightly convex, causing the bubbles to drift to the edge of the cup.

Fair weather cometh out of the north.

This saying comes from the Book of Job 37:22. Another version of this line states: "Out of the north cometh golden splendor." According to some commentaries, the proverb was derived from the observation that, in the Northern Hemisphere, northern winds scatter the clouds and clear the sky. A metaphorical view is that clouds symbolize sins that prevent humans from seeing God's true light.

A rainbow in the afternoon: Fair weather will be coming soon.

The appearance of a rainbow does not necessarily mean that there will be no more rain that day. A rainbow is formed by the diffraction of sunlight into its component wavelengths as the cloud cover begins to dissipate, permitting the sun to penetrate through falling rain. Rainbows always occur in the part of the sky opposite the sun—the western sky during the morning and eastern sky during the afternoon. Farmers would pay careful attention to where in the sky the rainbow appeared. If the rainbow shone in the area from which the storm originated, that would signal more rain. In the United Kingdom, for example, most storms come from the west, so a rainbow in the west might not be an auspicious sign after all. A rainbow in the east, however, would presage dry weather because it would mean that clouds from the stormy region of the sky had begun to break up as the depression that brought the storm began moving away.

Daily Bubble Forecast
If the bubbles in your coffee break up without forming in the middle or along the edge, then the baromatic pressure has reached a happy middle ground.

An early cuckoo heralds a fine summer.

According to English tradition, in order to qualify as "early," the cuckoo had to make itself known before St. George's Day, which falls on April 23. That would mean that you could depend on a good summer. The cuckoo is found throughout Britain and Ireland and has earned a reputation as a trickster: It lays its eggs in the nests of other species—some 50 in all in Britain alone. The first cuckoos are usually heard in mid-April, but peak migration takes place in late April or early May. In some years they can be heard as early as mid-March.

Quarter Past the Cuckoo!
The largest operating cuckoo clock in the world is in the ski resort town of Kimberley, British Columbia.

SETTING UP YOUR OWN WEATHER STATION

Basic weather stations are surprisingly easy and inexpensive to set up, and they can generate data with great sophistication, thanks to advanced meteorological software that is now commercially available. There are many organizations where amateurs and professional weather enthusiasts from all over the world exchange data and engage in online forums.

To assemble a weather station you will need:

- Thermometer, to measure temperature
- Barometer, to measure barometric pressure
- Hygrometer, to measure humidity
- Anemometer, to measure wind speed and direction
- Rain gauge, to measure precipitation
- Psychrometer, to measure the strength of evaporative cooling in the atmosphere
- Ceilometer, to measure cloud height
- Communication system, such as the Internet

To protect these instruments from the elements, you should place them in a simple structure. The most common shelter is called a Stevenson screen. Essentially a vented box, the Stevenson screen keeps the instruments ventilated without allowing them to become wet or dirty.

Barometers should not be placed in any areas that are affected by even small pressure changes caused by compression due to closing doors in small rooms, nor in areas that are exposed to vibration and rapid temperature changes. They should also be at least 3 feet (almost 1 m) above ground level.

Anemometers should be mounted 30 to 33 feet (10 to 12 m) above the ground in an area that allows free flow of air.

Thermometers should always be placed approximately 5 feet (1.5 m) above ground level, or 2 feet (0.6 m) above average maximum snow depth. They must be adequately ventilated, but kept away from direct sunlight.

Rain gauges should be level and close to the ground, and placed on flat terrain. They should be kept away from obstructions. Because of splashing, overflow, and other factors, they are usually the least accurate sensor in home units. Be sure that the gauge is always free of debris.

A veering wind will clear the sky;
a backing wind says storms are nigh.

The direction from which a wind comes is often less important than any change in its direction. That change is known as backing or veering. Whether a wind can be said to back or veer depends on the position of the observer at the time a storm is approaching or passing. If you point to where the wind came from and then at where the wind is currently blowing and your arm moves clockwise, then the storm has veered. If, on the other hand, you move your arm counterclockwise, then the storm has backed. A backing wind will bring rain. A convenient mnemonic device is: Veering is clearing.

A cow with its tail to the west makes weather the best.

This New England saying contains a degree of truth. Animals typically graze with their tails to the wind. This way, if a predator is approaching, its scent will travel and the animal will be alerted to its presence. Since most storms tend to follow a west-to-east course, an easterly wind is a rain wind and a westerly wind is a fair wind. Some folklore holds that the saying is correct only if at least half a dozen cows are grazing together in a field. They all have to be lying down together, too—it will not work if some are standing and some are lying down.

If fleecy white clouds cover the heavenly way,
no rain should mar your plans that day.

These fleecy white clouds are cumulus clouds, the billowy, cotton candy-like clouds. They are at least as tall as they are wide, and they develop on sunny days from packets of rising air. That gives them their reputation as fair-weather clouds. The base of each cloud is flat, and its top resembles a rounded tower. Some cumulus clouds can assume a towering appearance, in which case they are called congestus or towering cumulus. These clouds can build up into higher reaches of the atmosphere, and if they do so, they can develop into cumulonimbus clouds. Cumulonimbus clouds are tall, deep, and dark clouds that are formed from rapidly rising and sinking air currents—with warm air being lifted from the ground and cooler air funneling down in the opposite direction. These conditions promote violent thunderstorms that spawn fierce winds and tornadoes.

THE ATMOSPHERE

Earth's atmosphere consists of five layers: the troposphere, stratosphere, mesosphere, thermosphere, and exosphere. The layers are based on temperature changes at varying elevations. The troposphere is closest to Earth and is often referred to as "the weather zone" because that's where weather "happens." It extends approximately 5 to 9 miles (8 to 14.5 km) into the sky, just below the peak of Mount Everest. The troposphere is the shell in which life flourishes; in the atmospheric layers above it, the conditions are hostile to life. But without those layers, the Earth would be exposed to harmful radiation from the sun, not to mention chunks of rock (such as comets or meteors) and fragments of space junk from dead satellites.

The next layer, the stratosphere, extends to 31 miles (50 km) above Earth's surface, contains the protective ozone layer. The mesosphere extends to 53 miles (85 km) above Earth's surface and is succeeded by the thermosphere, which extends about 372 miles (600 km) above the surface of the Earth. This is known as the upper atmosphere.

Within the upper thermosphere is the ionosphere, which is ionized by solar radiation. Temperatures in this layer can get extremely hot—above an estimated 2732°F (1500°C). The last layer, the exosphere, serves as a kind of transition zone between Earth and space.

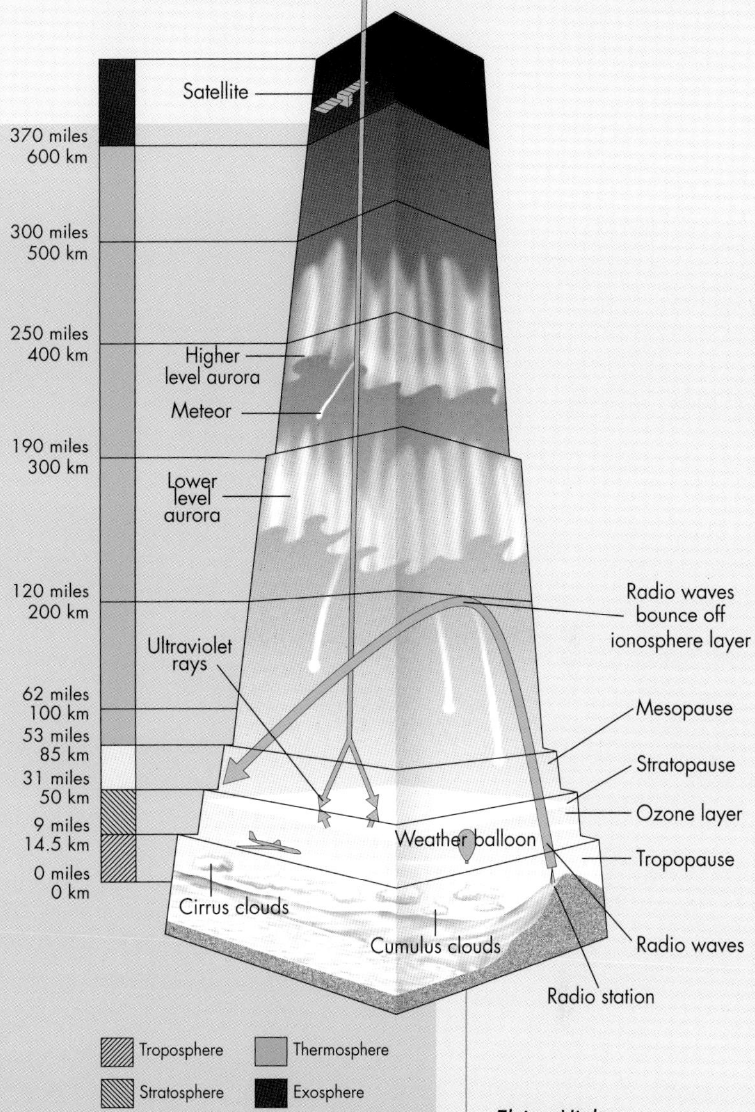

Flying High

Jet pilots prefer to fly in the stratosphere because conditions there are more stable than in the troposphere. The thermosphere above it is the region in which satellites and space shuttles orbit.

Open crocus, warm weather; closed crocus, cold weather.

This saying is similar to many that look to flora as an indicator of fair or foul weather. But the saying has only limited validity: Whether a crocus opens or closes is contingent not only on weather conditions, present or future, but on what type of crocus it is and where it is growing. Once *Crocus tournefortii* opens, for instance, it remains open. Maryland gardner Jim McKenney calls them "the working person's crocus" because "they will be waiting for you with, if not open arms, at least open tepals when you finally get home." Typically, crocuses are closed up tight when temperatures cool and daylight fades. If they are brought indoors and exposed to warmth, however, they will open. In that case the plant is responding not to future weather conditions, but simply to the radiator.

THE OZONE HOLE

Ozone, a form of oxygen, is mainly found in the stratosphere. It is especially important because it shields human life from dangerous ultraviolet rays of the sun that can cause skin cancer and eye damage. In 1985, British scientists Joesph Farman, Brian Gardiner, and Jonathan Shanklin of the British Antarctic Survey made a startling discovery: The ozone layer over Antarctica had developed a gap, a development that seemed to recur in spring in the Southern Hemisphere (September through November). The hole tended to close again because of changing conditions (much lower temperatures and higher winds). The culprits were identified as chloroflourocarbons (CFCs) and bromofluorocarbons, highly stable compounds that are capable of surviving intact as they rise into the stratosphere; there the ultraviolet radiation breaks them down.

The resulting chemical reactions caused the depletion of the ozone layer, opening a large hole of up to 50 percent over the Antarctic. While the ozone hole is seasonal—it shrinks in fall and winter—it would expand alarmingly each time it reappeared in spring. In September 1987, negotiators from around the world signed an accord—the Montreal Protocol on Substances That Deplete the Ozone Layer—to limit the production and use of manmade chlorines that contributed to the problem. As a result of the protocol, CFCs have decreased dramatically.

When ants travel in a straight line, expect rain; when they scatter, expect fair weather.

There is some evidence, based on both traditional observation and scientific studies, that the collective behavior of ants can have predictive value in regard to the weather. The indigenous people of Australia would pay heed to ants when they built walls around their nests, since it meant that heavy downpours would follow. But in general, ants react to weather conditions rather than forecast them. Studies have shown, for instance, that fire ants and rainfall have an opposite relationship. The more rain, the better it is for the ants. Fire ant mounds will rise from the ground in response to elevated soil moisture. The moist environment is congenial for fire ants to breed with even more ardor. In winter rainstorms or in summer droughts, however, the fire ants will look for shelter indoors. Ant invasions are not forestalled by pesticides so much as what the weather is like outside.

If Michaelmas sets clear and bright, it will stay dry with frost at night.

By the end of September—Michaelmas falls on September 29—farmers would generally be in the midst of plowing fields used for growing cereals and root crops. But this task could be done only if the weather was dry. The prevalence of calm conditions, cooler nights, and light winds, which result from a high-pressure system, indicates that conditions will remain dry for the next several days. Those same conditions also promote the formation of frost.

The higher the clouds, the better the weather.

It would be a mistake to invest too much trust in this saying: While some high clouds presage fair weather, several types of high clouds can bring violent storms. There are three basic types of high clouds: cirrus, cirrostratus, and cirrocumulus. White cirrus clouds are the most common, and they are the ones that are associated with fair weather. Composed of ice, they consist of long, thin, wispy streamers that trail in the wind and give them their popular name of mares' tails. The movement of cirrus clouds across the sky will tell you which direction weather is approaching from. The appearance of cirrus clouds usually means that a change in weather is likely within 24 hours. When you see cirrostratus clouds—easily identified as sheetlike, thin clouds that can cover the entire sky—you can expect that rain or snow will follow within 12 to 24 hours. Cirrocumulus clouds are small, rounded, and puffy, usually appearing in long rows. White or gray, they are usually seen in the winter and indicate fair but cold weather. In the tropics, however, they can be a harbinger of an approaching hurricane.

WEATHER LORE IN THE SOUTH SEA ISLANDS

Before setting out on voyages across the Pacific, ancient Hawaiian, Polynesian, and Micronesian peoples would examine natural portents for signs of what the weather would be like at sea. They would observe the swells, the color of the sky, and the shapes, colors, and movements of clouds to forecast weather.

At night they would study the twinkling of the stars (a sign of wind direction). If a halo surrounded the moon and more than 10 stars could be seen through it, they believed that rain and wind were on the way. A double halo portended a torrential downpour.

If cats lick themselves, fair weather.

During fair weather, relative humidity is low. Under conditions of high pressure, dry air sinks from above; relative humidity is low, and cat hair becomes a better insulator. As a result, static electricity can build up. Humidity is normally lower in the winter, and heating the house reduces the humidity further still. (A desert climate typically has very low relative humidity.) Cats, like humans, do not like being shocked when they come into contact with other objects. In addition, their hair loses electrons easily, so cats become positively charged, making them more vulnerable to charges. By licking themselves, cats are increasing the amount of moisture in their fur, making it more conductive so that the charge can "leak" off them. That's why many cats don't like to be petted during cold winter weather, because the accumulated charge can cause small sparks. That puts them in a bad mood, and it does not make their owners any happier, either.

If cocks crow during a downpour, it will be fine before night.

Cocks (or roosters) are best known for greeting the dawn (and waking people up), so they have earned a reputation as the "sun bird" and as a harbinger of dawn and light. Not surprisingly, various religions have embraced them as symbols. The cock was supposedly the first animal to proclaim the birth of Jesus, for example. However, the cock could inspire dread if it crowed at any time other than dawn; in certain cultures it was considered an omen of death. Cocks have also been used to forecast the weather. In England, if a cock sits on a fence or gate and begins to crow, it is said to mean that it will rain the next day. There is no evidence, however, to support that its crowing during a rainstorm will ensure clearing skies by nightfall.

When the sun sets bright and clear, an easterly wind you need not fear.

Easterly winds are dry winds that typically bring colder air and clear skies. (Winds are named for the direction from which they come, not where they are blowing.) The Polar easterlies, as they are known, blow from east to west. Easterlies, like westerlies and other large-scale winds, push and pull weather with them, so they often provide an indication of approaching weather conditions. In winter, easterlies that reach Britain, for instance, bring frigid air from Russia. Two of the coldest winters on record in the British Isles—the winters of 1947 to '48 and 1962 to '63—are blamed on these winds. As dry as they are, these winds can also collect moisture when they travel over bodies of water, so their arrival is sometimes typified by bands of low clouds. Some local easterly winds, on the other hand, such as the Levant that blows over the Mediterranean, will bring mild, moist weather to Gibraltar, Spain, and Africa.

CATS AND DOGS AND WEATHER

There is no question that cats—more than dogs—have served the purpose of weather diviners for centuries.

- When it rains heavily in Britain and North America, you might say that it is "raining cats and dogs."
- English mariners referred to a very frisky cat as a cat with a gale of wind in its tail.
- A northwest wind that brings stormy weather to the Harz Mountains in Germany is called a cat's nose.
- Both cats and dogs were attendants of the mighty Odin, the Norse god of storms. The cat symbolized downpours, while the dog represented blustery winds.
- Old German pictures depict the wind as the "head of a dog or wolf."
- Rain is supposed to be more likely when a cat washes its ears or licks its fur.
- If a sailor was about to embark on a voyage and his cat began to mew loudly, he might want to reconsider, because it meant the journey would be difficult. A playful cat meant smooth sailing because the winds would be helping the ship. If, however, the cat simply dropped off to sleep with all four of its paws tucked under its body, its owner would be advised to stay put because it meant that bad weather was coming.

Red sky at night, sailor's delight.

The origin of this common saying has been attributed to an English nursery rhyme, although many regions and cultures have their own variations. The expression changes when it applies to land: The red sky at night is also a shepherd's delight. In either case, a red sky at night meant clearing weather the following morning, while its reverse—a red sky in the morning—meant bad weather that day ("Red sky in morning, sailors take warning"). Folklore historians have traced the saying to the New Testament, specifically to Matthew 16:2–3: "He (Jesus) answered and said unto them: 'When it is evening, ye say, It will be fair weather: for the sky is red. And in the morning, it will be foul weather today: for the sky is red and lowering. O ye hypocrites, ye can discern the face of the sky; but can ye not discern the signs of the times?'"

Red Sky at Night...
One of the most popular weather lore sayings, this was also referenced by Shakespeare in Venus and Adonis*: "Like a red morn, that ever yet betoken'd wreck to the seaman—sorrow to shepherds."*

Whatever its true origins, the saying does have some scientific basis and can serve as a reasonably reliable indicator of weather—at least at the mid-latitudes in the Northern Hemisphere, where storm systems generally follow the jet stream from west to east and are pushed ahead by the westerly trade winds. At sunset and sunrise the sun is low on the horizon; as a result, its rays have to pass through a thicker layer of atmosphere than it does when it is higher in the sky. The atmosphere scatters out most of the shorter wavelengths of the visible spectrum—the greens, blues, and violets—which allows the reds that have longer wavelengths, to predominate. It is important to keep in mind that the atmosphere is not just composed of clouds full of water vapor; it also may contain dust particles and pollutants, all of which can impede some wavelengths and amplify others. The deeper and more fiery the reds, the more moisture, dust, and pollutants there are in the sky.

A red sky in the evening often means the setting sun in the western sky is shining on the underside of moisture-laden clouds in the opposite direction. The sun's rays need a direct path to illuminate the clouds, which in turn indicates clearing conditions. In other words, there are fewer clouds to get in the way. As the weather continues to move east (carried by the westerly wind), precipitation

CAN PLANTS FORECAST WEATHER?

Traditionally, plants have served as meteorological tools as much as animals have. It is believed, for example, that when leaves show their backs, it will rain. There is some scientific basis for this: As trees grow, their leaves assume a certain pattern that is heavily influenced by the prevailing wind. A storm wind is an anomalous wind, insofar as it is not the prevailing one, and so the leaves will be ruffled in such a way as to display their lighter-colored undersides. Marigolds, pimpernels, crocuses, morning glory, and many other plants and flowers will also close or open, depending on the degree of moisture in the air. Storms are not the only cause; humidity and temperature can also have the same impact.

Scientists are also beginning to reexamine the usefulness of plants for meteorological purposes. Researchers at Purdue University in Indiana are studying whether the amount of moisture that plants emit during photosynthesis could be "the local trigger that trips fronts into violent weather." (Photosynthesis is the process by which plants employ solar energy to convert carbon dioxide and water into carbohydrates.) These researchers are simply following in the footsteps of plant biologists who have used models of photosynthesis to track environmental changes.

will often move off as well. Under those circumstances, at sunrise the following morning, the eastern horizon is likely to be gray. This means that the storm clouds are already to the east of the area and you can expect fair weather to prevail as the day wears on. There are certain conditions, however. For one thing, regions of precipitation do not invariably move west to east. For another, clouds aren't always associated with approaching or departing storm systems. Finally, it is always possible that another storm system might move in later in the day, dashing expectations.

Marigolds

Like other flowers, marigolds are credited with the ability to forecast weather because their leaves will open or close depending on the amount of moisture in the air.

A native of Mexico, marigolds are one of the most popular bedding plants in the United States. They are popular warm season annuals and come in a variety of colors. They require a sunny location, fertile soil, and moderate water to flourish.

Heat Waves

Heat waves have long been feared as much as icy winter weather—and for good reason. In the past, persistent days of heat could not only ruin crops, but they were associated with the spread of infectious diseases that both livestock and humans were susceptible to. Summer solstice festivals, while celebrating the arrival of the summer, were also intended to ward off disease. Even in recent times, heat waves have caused the deaths of thousands of people across the world from dehydration.

The wind shifts around to the south or southwest, expect warm weather.

Winds from the south often bring hot, dry weather—sometimes too hot and dry, as Shakespeare noted in *The Tempest:* "A southwest blow on ye, And blister ye all over." The Bard might have been referring to the Khamsin, a hot wind that blows out of the interior of Africa over Egypt and into the eastern Mediterranean. It carries a great deal of dust; it is a source of dread for those who have to endure it, especially because the deposit of dust it leaves in its wake can destroy crops. The Leveche, another local wind, blows in southeast Spain and often signals the approach of a depression associated with storms. To predict future weather systems it is not enough to know where the wind is blowing from; a change in wind direction may be just as or more important. A wind that shifts around to the south or southwest is likely to bring warmer temperatures; but when winds are already blowing out of the south-southwest in warmer weather, it indicates that a cold front is approaching. Once the front passes through, the winds will usually shift around to the west-northwest, setting the stage for an inflow of colder air.

As the days begin to shorten, the heat begins to scorch them.

This saying refers to summer weather, when the days are growing shorter after the summer solstice. August has a reputation for especially hot weather, never more so than in recent years, when the month has included unprecedented heat waves.

FATA MORGANA

A Fata Morgana is a superior mirage (*see* How Mirages Work, facing page). They may be seen toward the horizon above cold water or ice as displaced multiple images stacked above the object's real location. The effect on cliffs and icebergs, for example, may give them a castlelike appearance.

The term *Fata Morgana* (Italian for "Fairy Morgan") has its origins in Morgana le Fay, the mythical shape-shifting enchantress, who lived in a castle under the sea and whose witchcraft was thought to cause the mirage. Sailors in search of safe harbor would sometimes confuse the mirage for a real castle on the ocean's surface.

In August 2003 a heat wave struck Europe with disastrous consequences, killing some 35,000 people. The United States and Canada also suffered through fierce heat. In fact, that August was the hottest on record. The Earth Policy Institute (EPI), a think tank based in Washington, D.C., declared: "Though heat waves rarely are given adequate attention, they claim more lives each year than floods, tornadoes, and hurricanes combined. Heat waves are a silent killer, mostly affecting the elderly, the very young, or the chronically ill." The August heat claimed about 7,000 lives in Germany, nearly 4,200 lives in both Spain and Italy, over 2,000 people in the United Kingdom (which recorded temperatures over 100°F (37.8°C) for the first time). In France the total reached nearly 15,000; many of the victims were elderly people who lived in houses without any air conditioning. The EPI projects more extreme weather and higher temperatures in summer as the century goes on, with the average world temperature projected to climb from 34.5° to 42.4°F (from 1.4° to 5.8°C) by the year 2100.

Desert Mirage

A type of superior mirage (see How Mirages Work, below) known as looming occurs in deserts when a false oasis appears. It is the displaced image of a real, but distant, oasis beyond the horizon. Looming also occurs at sunrise: When you see an image of the sun's disk falling below the horizon, it has in fact already fallen.

Swallows high, staying dry;
swallows low, wet will blow.

This saying has some credibility. Birds will fly higher when conditions are calm and winds are light because there is less resistance. But they have another motive, too. When good weather prevails, the insects that the birds feed on are carried up high on warm thermal currents rising from the ground, meaning that the birds are assured a source of food without having to dive down for it.

HOW MIRAGES WORK

Mirages occur when there is a sudden change in the temperature, and therefore the density, of adjacent atmospheric layers. The differing densities cause light waves to slow down in cold layers and speed up in hot layers, and this results in the light wave bending. As the light bends, it may produce a displaced image of an object, and the image may be inverted and distorted. Inferior mirages, in which the displaced image appears below the location of the real object, often appear on hot asphalt roads or deserts, when an image of the sky appears as a dark, shimmering pool on the road or sand, or an inverted image of a car appears beneath the real car. In a superior mirage, the displaced image appears above the object (*see* Fata Morgana, facing page). Both the inferior and superior image can look like a simple reflection.

Weather on the Web

Spiders have long been associated with myth and folklore. However, there is no evidence that a spider spinning its web indicates that a fine day lies in store or that in advance of a storm spiders abandon their webs. Spiders seek shelter in cooler weather, which is why infestations of houses are more likely to take place in winter.

If spiders are many and spinning their webs, the spell will soon be very dry.

It is doubtful that a proliferation of spiders forecasts warm spells, so much as it is a response to rising temperatures. Warmer weather will attract more spiders and was the likely explanation for an influx of immigrant spiders from mainland Europe that turned up in the south of England during the summer of 2006, most likely on cross-channel shipments of fruit. Their appearance was hardly a matter of chance: Warmer weather provides more favorable conditions for insects to breed. So there was more food for the spiders—mostly false widows and wasp spiders—to eat.

Chinook winds can cause a sharp increase in the number of migraine headaches

The Chinook is a katabatic wind, derived from the Greek word *katabatikos*, meaning "going downhill." Chinooks, like foehns and Santa Ana winds, flow up the crest of a mountain, then pick up speed as they move downhill, bringing hot, dry air when they reach the ground on the other side of the mountain. The Chinook blows in Canada and in the western United States. These winds are so strong that they are capable of warming the air in a matter of minutes. Early in the morning on January 22, 1943, for example, the temperature in Spearfish, South Dakota was -4°F (-20°C). Within 2 minutes of the Chinook blowing in, the temperature had soared to 45°F (7°C). The incredible rise in 2 minutes was a world record. A couple of hours later the temperature was in the mid-fifties. But once the Chinook had blown off, the temperature dropped to -12°F (-24°C) in just 27 minutes.

Chinooks are most prevalent over the Canadian province of Alberta where they occur on an average of 30 to 35 days a year, which is why southern Alberta is known as Chinook Country. In some parts of the province, winds can reach hurricane force, and winds as high as 107 m.p.h. (172 kph) have been recorded.

In winter these winds can trap pollutants in the air and produce an inversion fog—a fog caused by a rise of temperature with an increase in height—which is believed to increase irritability and sleeplessness. It appears that air pressure has the effect of aggravating migraines as well as pain in the joints. "The weather factor is a very common trigger for many migraine sufferers," said Dr. George Urban, associate director of the Diamond Headache Clinic in Chicago. "Usually, it is during or before [a decline] in barometric pressure."

EL NIÑO

El Niño is an unpredictable warming of the Pacific Ocean. It has a great impact on global climate, producing drier conditions in southeast Asia, rainier weather in South America, and warmer winters in the United States. It is caused by a weak, warm current in the Pacific and gets its name—Spanish for "little boy," after the Christ Child—because the pattern is usually seen around Christmastime.

El Niño tends to occur every three to seven years. El Niño develops as trade winds in the tropical Pacific drive the surface waters to the west; once they reach the coastal waters of Peru and Ecuador, they begin to replace the cooler, nutrient-rich waters. Because the warmer water imported from the tropics carries few nutrients, it leads to a significant reduction in marine life. At the same time rainfall follows the warm water eastward—toward Peru—and leaves the western Pacific with less precipitation. That accounts for why El Niño is the cause of increased rainfall and flooding across the southern United States and Peru. It causes droughts in the western Pacific, which make Australia vulnerable to devastating fires. Severe El Niño events have been responsible for thousands of deaths around the world, have destroyed homes, and have caused billions of dollars in damage.

The less frequent La Niña ("Little Girl"), by contrast, occurs when surface temperatures in the eastern tropical Pacific become cooler than normal and produces warmer winter temperatures than normal in the U.S. southeast and southwest and below-normal temperatures in the northern United States—just the opposite effect from El Niño.

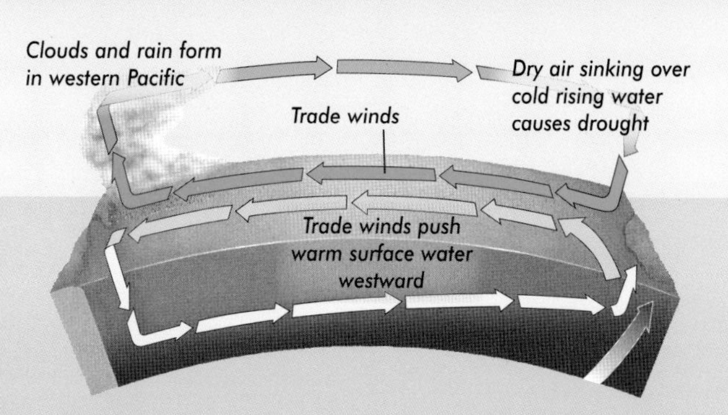

La Niña

This occurs when waters in the eastern Pacific begin to cool, causing air temperatures to drop. Trade winds push warmer waters west, resulting in moister, warmer weather in coastal regions in the western Pacific.

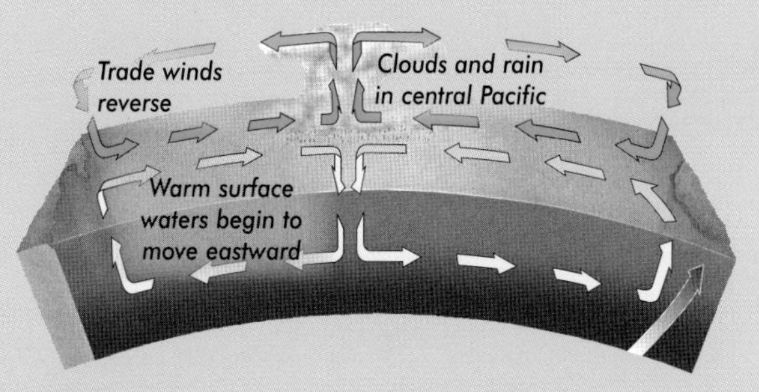

Transition

When the trade winds reverse direction, La Niña gives way to a better known phenomenon called El Niño.

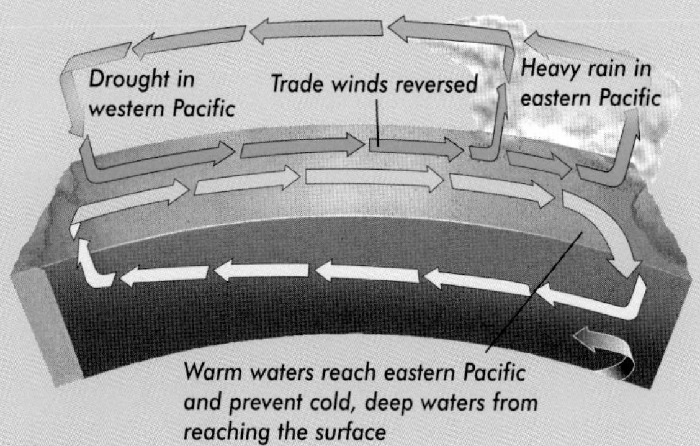

El Niño

Weather throughout the world is affected by this significant warming trend and rain in the eastern Pacific. It can persist for months and occasionally up to a year.

Cricket

Crickets, like other insects, are supposedly more active before a rainstorm.

Crickets chirp faster when it is warm and slower when it is cold.

Crickets can indeed serve as thermometers. Tradition says that if you count the cricket's chirps for 14 seconds and then add 40, you will obtain the temperature in Fahrenheit at the cricket's location.

If frogs make a noise at the time of cold rain, warm dry weather will follow.

Frogs figure prominently in folklore and mythology. And a lot of that folklore is based on the observation that the amphibians do in fact make noise before rain. Some Native Australians and Native Americans believed that frogs announced or even summoned the rain. In India, frogs were associated with thunder. In Sanskrit, which is still used as a ceremonial language in India, the word *frog* meant "cloud." Scientists have taken over where folklorists have left off, trying to find out what effect weather has on frogs, if not vice versa. They found that rainfall strongly affected the initiation of calling in at least one frog species—*Austrochaperina robusta* males—but that the amount of humidity as well as rainfall influenced whether they continued to call. The males of another species, *Calcarius omatus*, were much more strongly affected by humidity than by rain. The researchers noted that calling levels within species "did not appear to be caused by common responses to the weather variables." While weather did seem to play a role in the initiation of the amphibians' calling behavior, it was also influenced even more by social interactions among the species.

Locusts sing when the air is hot and dry.

In Books 1 and 2 of Joel it is written: "The fields are ruined, the ground is dried up, the grain is destroyed, the new wine is dried up, the oil fails. Despair you farmers, wail you vine growers; grieve for the wheat and the barley, because the harvest of the field is destroyed...the land is as Garden of Eden before them [the locusts], and behind them, a desolate wilderness, yea, and nothing shall escape

WARM FRONTS

Warm fronts bring air that is warmer and moister than the air ahead of it. They are much more sluggish than cold fronts. These fronts are defined as the transition zone where a warm air mass is replacing a cold air mass. Sometimes the cold air mass, which is denser than the warm air, will refuse to give way, a situation that can lead to several days of wet weather—an effect known as cold air damming.

Although they are capable of causing thunderstorms, warm fronts are typically less violent than cold fronts and more frequently produce light to moderate continuous rain to the north of the front, as the warm air, driven by low-level southerly winds, rises over cooler, denser air. This lifting results in cloud formations and precipitation. On a weather map, circles on the red line demarcating the boundary between the two fronts indicate the direction the warm air is moving toward. The boundary can extend for hundreds of miles over the cold air.

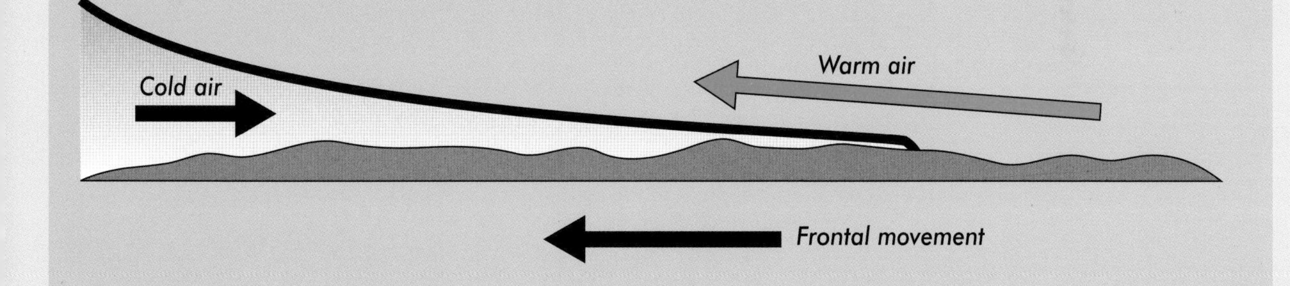

them...." The plague of locusts vividly described by the Old Testament is not simply a historical episode. Swarms of locusts continue to devastate various regions of the world, most recently in sub-Saharan Africa, where they can devour crops and threaten the livelihood of hundreds of thousands of people.

Locusts are especially prone to appear in large numbers in hot and humid conditions. In 2004 locust swarms had already made their appearance in parts of Cape Verde, Mauritania, Mali, Senegal, and Niger, wreaking havoc on the summer planting season. Although such swarms are fairly frequent in West Africa, the locusts in the summer of 2004 were considered the most serious to hit that part of the world since the 1987 to 1989 infestation. Locusts breed in hot, humid weather and move with the wind. However, heavy rains are conducive to breeding. In Australia the national meteorological service issues a "locust weather forecast," but "only when conditions favor the large-scale southward movement of locusts from northern South Australia."

HUMIDITY

Humidity has been a particular focus of meteorologists, just as it has to folklore—not only because it can make for uncomfortable weather, but because it is a measure of the amount of moisture in the air. That in turn offers a good indication of an approaching weather system, such as a thunderstorm. Sailors, who like farmers were preoccupied by weather, relied on ropes to give them an idea of what to expect in the waters that they planned to navigate.

When rope twists, forget your haying.

The rope refered to in the saying—hemp—was traditionally employed by farmers as a hygrometer, a weather instrument used to measure humidity in the air. The hemp was exposed to circulating air and would get kinks in it with increasing humidity. Some farmers also observed the softness of tobacco to measure the amount of water in the air.

Curls that kink and cords that bind.

Sailors relied on rope and seaweed to measure the humidity in the air. Rope served the same purpose as a hygrometer. The more humidity in the air, the more the rope would kink. The larger amount of water in the air, the greater the likelihood that a storm system was approaching, which would bring precipitation and heavy wind.

Pimpernel, pimpernel, tell me true
whether the weather be fine or no;
no heart can think, no tongue can tell,
the virtues of the pimpernel.

Plants and certain fungi such as rainstars have traditionally been used to forecast wet and dry weather. Farmers observed that the bog (or scarlet) pimpernel, wild indigo, chickweed, dandelions, bindweeds, clovers, and tulips all fold up before it rains when moisture is abundant in the air. When the atmosphere reaches about

WHAT IS HUMIDITY?
Basically, humidity is the amount of water vapor in the air. Water accumulates in the air as a result of evaporation; water ceases to be a liquid and turns into a gas. In some conditions, such as a desert, there is little water on the ground and so the humidity will generally be low.

When too much water vapor has collected, it reaches a saturation point: Clouds form, and after condensing and cooling in the upper atmosphere, the water returns to Earth as precipitation. There are three types of humidity: relative, absolute, and specific.

MEASURING HUMIDITY

Humidity is typically measured by a hygrometer. The device measures relative humidity by first determining absolute humidity—the mass of water vapor in a given volume of air. Relative humidity is the amount of moisture present in the air divided by the amount of moisture the air can hold. The most common hygrometer is the dry- and wet-bulb psychrometer. One bulb has a moist cotton or linen wick wrapped around it. As the water in the cloth evaporates, the wick absorbs heat from the thermometer bulb, causing the wet thermometer reading to drop because the loss of heat is cooling the air. Relative humidity is determined by calculating the difference between the reading from the dry thermometer and the reading from the wet thermometer.

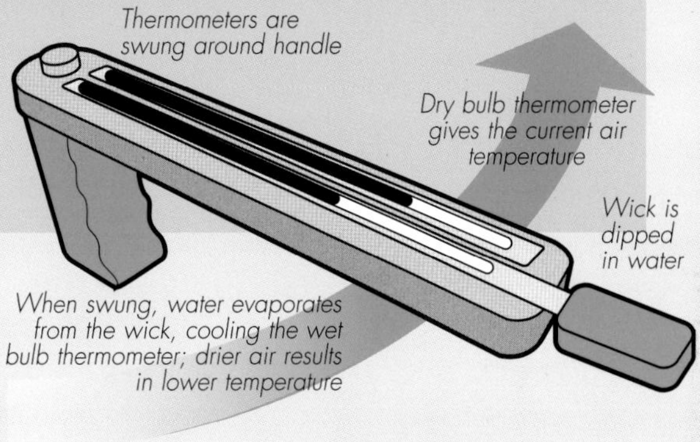

Thermometers are swung around handle

Dry bulb thermometer gives the current air temperature

Wick is dipped in water

When swung, water evaporates from the wick, cooling the wet bulb thermometer; drier air results in lower temperature

Psychrometer

This type of hygrometer consists of a wet bulb and a dry bulb thermometer. This type is known as a sling psychrometer because the thermometers are attached to the end of a sling or chain, which once whirled, will ensure proper ventilation and more accurate readings.

80 percent humidity, the bog pimpernel closes, which accounts for this saying. The pimpernel is also known as a Shepherd's Barometer or a Poor Man's Weatherglass, attesting to its reputation as a botanical forecaster. The pimpernel is widely distributed in all temperate regions in both hemispheres. Its egg-shaped leaves are arranged in pairs, and regardless of the angle of the stem, always face the light. The petals are very sensitive, so much so that the flowers will close even if the skies cloud up and it simply looks like rain. Even in fair weather the flowers are late risers, never opening until 8 or 9 in the morning. By 3 in the afternoon, their work is done, and they close up for the day. So, given the abbreviated hours they keep, their value as a prognosticator is still rather limited.

The leaves of the May tree bear up so that the under side may be seen.

Many plants and flowers—including pimpernels, marigolds, and crocuses—will open or close, depending on the amount of moisture in the air; but the same rule does not necessarily apply to the leaves of trees. Tree leaves will show their lighter-colored undersides before a storm, but this effect may be due less to the humidity than to the way the wind is blowing. As leaves grow, they form a particular pattern that is mostly governed by the prevailing winds. However, winds that blow in advance of a storm—which are different from the prevailing winds—will ruffle the leaves in a way that exposes their undersides.

Droughts

"The clouds that carried the rain sailed high above, not seeming to notice the suffering of Africa. No fruits, no fodder, and hardly anything to drink." Thus begins a myth from South Africa. Fortunately, the rain god finally pays heed to the parched continent when he becomes attracted to a young woman named Savuri. "Her skin looked like shining wet rock, her hair was as dark as dew-moist berries, and the rain desired her." Folklore is not just concerned with predicting impending drought but with ways to stop it. Sometimes a beautiful woman is as promising a solution as any.

One day's rain drowns out seven weeks of drought.

This saying from Germany is problematic; soil that has been parched for weeks or months is unable to absorb a great deal of rain all at once—quite the opposite. The excess water is more likely to run off and cause flooding. It is much better to have modest rainfalls over an extended period to replenish the lost moisture and allow the

DROUGHT LEGENDS

Legends about droughts—and ways of ending them—have inevitably arisen among people who have seen their crops wither and die under perpetually cloudless skies. The Zuni Indians, for instance, attributed droughts to the malevolence of a monster called Cloud Eater. He stood as tall as a mountain peak and had an insatiable appetite for clouds, leaving none in the sky to produce rain. The Zunis would hunt him far and wide to stop his depredations so that they would have rain, but no one knew where to find him.

In ancient India, it was said that a dragon stood guard over the clouds to hoard the rain and prevent it from falling to Earth. People would appeal to the storm god to lure the dragon away and allow rain to fall. In the U.S. Great Plains, a legendary figure named Febold Feboldson took matters into his own hands when drought befell the land. "He liked his fishin' right enough, and there was no fishin' to be had in that drought. So he sat down and thought up a way to bust that there drought." He proceeded to build bonfires around all the lakes in the region to evaporate their water. The water eventually condensed into clouds—so many of them that they "bumped into one another"—and rain began to fall.

earth to recover. When the ground is extremely dry, it makes a drought worse. That is because under normal conditions, some of the sun's energy is used to evaporate moisture in the ground or in vegetation. With no moisture in the soil to evaporate, it leaves more solar energy available to heat up the atmosphere even further.

Dog days are when dogs go mad.

The Dog Days, so named after Sirius, the Dog star (see box on page 30), dominate the skies in midsummer in the Northern Hemisphere. In the summer, Sirius rises and sets with the sun. It was believed that the heat of the star together with the heat of the sun made the weather on the earth especially hot and sultry. Today we know that that it is the tilt of our planet that is responsible for seasonal changes. It was a period in which, owing to the paucity of rainfall, droughts were most likely to occur, bringing with them famine and disease. Just how perilous this period was reputed to be is illustrated by a passage from a book published in 1729 called *The Husbandman's Practice:* "All this time the Heat of the Sun is so fervent and violent that Men's bodies at Midnight sweat as at Midday: and if they be hurt, they be more sick than at any other time, yea very near Dead. In these days all venomous serpents creep, fly, and gender, so that many are annoyed thereby; in these times a Fire is good night and day, and wholesome, seeth well your meals and take heed of feeding violently."

In ancient Rome, July's heat was associated with diseases. When this star rises in the morning, ancient authors warned, the sea would boil, wine would turn sour, people would grow hysterical, and animals would become sluggish, with the exception of dogs—they would go mad. On July 3, the Romans would sacrifice a brown dog to placate the dog god Canicula. The Incans practiced a similar custom: In times of drought they would tie up black dogs and let them die of thirst in hope of appeasing Ilyap, the god of thunder and lightning. The belief that dogs were susceptible to madness during this time persisted for many centuries: At one time in England, some magistrates would order dogs to be muzzled from the beginning of July.

EXTREME DROUGHTS

Cape Verde suffered some of the most severe droughts in the nineteenth century; three successive droughts resulted in as many as 100,000 deaths and led to mass migration. In 1921 to '22, a drought in the Ukraine and Volga claimed the lives of at least 5 million people, many of whom perished from famine. The worst drought in recorded history took place in 1936 in China's Szechuan Province, in which 5 million people perished and over 34 million farmers were displaced.

After great droughts come great rains.

It is true that in many parts of the world, dry periods are followed by prolonged wet periods. Until recently, when climate change began to disrupt patterns, East Africa would experience rainfall in the autumn and then a heavier rainfall in the winter before the onset of dry weather that usually lasted through the spring and summer. If, however, a drought is episodic or due to manmade factors such as deforestation, then drought-like conditions can persist for months or even years.

Dust rising in dry weather is a sign of approaching change.

Dust rising in dry weather will probably portend change, but for the worse. Sand and dust storms have occurred for centuries; Chinese records speak of sandstorms occurring around A.D. 960, when a yellow dust covered the sky and laid waste to the fields. The number of such storms has been increasing in recent years; China, for example, experienced 13 storms in the 1970s, and 20 in the 1990s. Sandstorms and dust storms are global phenomena because of the action of air currents. For example, in 2000 some 8 million tons of sand originating in the Sahara was deposited in Puerto Rico. Cool, moist winds from the Mediterranean are to blame for many of these storms: When they move in over the Sahara, they force the hotter, lighter air to rise higher into the atmosphere, and as it does so it collects a lot of dust. Once in the upper atmosphere, the dust-laden air is at the whim of prevailing winds that push it in a westerly direction—toward the United States and Caribbean. It only takes about a week for sand to travel across the Atlantic to the U.S. eastern seaboard.

Dust storms can occur all over the world. While many of these storms occur naturally, humans share a great deal of the responsibility, too: Overgrazing, deforestation, and the abuse of water resources have contributed to the problem. This has never been more vividly—or catastrophically—illustrated than during the so-called "dirty days" in the U.S. Great Plains from 1934 to 1936, when the region was plagued by a series of record droughts. Once-arable land was transformed into virtual deserts. If there is any one cause of the Dust Bowl, as it is known to history, it was greed. Farmers overplanted and overharvested, trying to take advantage of high prices for crops; but when the bottom dropped out of the market, they abandoned their farms in great numbers, leaving the land unplanted and unplowed, making it especially vulnerable to dust storms when the droughts came.

WHY DROUGHTS OCCUR

An absence of rain in a particular area over an extended time can bring about drought conditions, but other factors may come into play as well. A drought is defined as "a period of abnormally dry weather sufficiently prolonged for the lack of water to cause serious hydrologic imbalance in the affected area," according to the *Glossary of Meteorology*. Their severity depends on the extent of moisture depletion, the size of the area affected, and duration. There are four ways of assessing a drought. The meteorological method compares rainfall in a particular period to find how much less precipitation there is than normally would be expected. So a drought in one area might not be considered a drought in another where, for instance, precipitation is relatively low to begin with. The agricultural method defines a drought as one where insufficient moisture remains in the soil to sustain a particular crop. The hydrological method measures the surface and subsurface water supplies; if they fall below normal, then a drought is said to occur. The socioeconomic method relies on the impact of prolonged scarcity of precipitation and water on the population. A drought may occur when snowfall is abnormally low in winter, leaving less runoff to nourish the soil in spring. Overplanting, deforestation, and desertification also promote conditions that may lead to severe drought.

1. *Jet stream*
2. *Ridge of high pressure*
3. *Trough of low pressure*
4. *Low pressure area*
5. *Blocking high*
6. *Low pressure area*
7. *Jet stream snakes far north and splits into two streams—the southern becomes stronger and the northern dies out.*

Blocking Highs

High pressure systems known as "blocking highs," which disrupt the normal pattern of the jet stream, can produce clear skies and hot, dry weather in summers. Ordinarily the jet stream steers lows into a region, which can cause winds and precipitation. As the jet stream travels above Earth it creates a giant loop, dragging the lows after it. However, if the system is blocked by a high, the rain never arrives. If the blocking high lingers for a prolonged period, a drought can result.

4

FOG, RAIN, CLOUDS, AND FLOODS

According to an old Burmese legend, people used to take weather, quite literally, into their own hands. They would stage a contest in which one group—the rain contingent—would tug on a rope that extended up into the sky to bring down rain, while their opponents would try to bring down drought. Usually rain was allowed to win. There is probably no phenomenon that has produced more weather sayings than rain and its related weather events—clouds and floods.

London "Pea souper"

A bus inspector uses a flare to lead a London double-decker bus over a crossing in the fog in 1952.

FOG

Fog can be romantic and mysterious; fog can also be deadly. Many Londoners can remember suffering through "pea soupers," in which large numbers of people died from breathing toxic pollutants that were trapped by the fog. Fog has long been used in folklore to predict weather. The number of incidents of fog in summer, for example, was thought to predict the number of snowstorms in winter.

A summer fog for fair,
a winter fog for rain.
A fact most everywhere,
in valley or on plain.

Fog is formed when the air cools enough so that the water vapor in the air condenses. For fog to form on a summer night, the air must be cool, the sky clear, and winds light, so that the excess heat can be radiated into space. (Cloudy skies act like a blanket, keeping the heat in; that heat would prevent fog from developing.)

By contrast, winter fog is formed under two entirely different circumstances. The air above an ocean or large lake is typically more humid than the air over land. When the humid air moves over cold land, it will form fog and precipitation. In climates that are very cold, when the temperature drops substantially below freezing, ice fog might form. Ice fog is typically produced by water vapor emitted by automobiles, household furnaces, and industrial plants. The vapor immediately condenses in the frigid air and forms fog.

Fog goes a hoppin', rain comes a droppin'.
Fog in January makes a wet spring;
February fog means a frost in May.

If fog forms in cold winter weather, then warm, moist air must be involved. The humid air typically collects its moisture from a body of water; it then pulls the moisture along with it as it moves over land. As the air meets cooler temperatures, it may condense and precipitate out as fog. However, the saying applies to conditions in winter when conditions are wetter and temperatures are above average, which may set the stage for an especially rainy spring. The same conditions that can promote ground fog in winter can also prevail in late spring, or May—that is, cooler temperatures close to the ground and clear skies—so that frost may form on the ground early in the morning. But so many other factors can come into play, including air currents or storms, that the predictive value of the saying is limited.

Legendary Dartmoor

Dartmoor has long been host to headless horsemen, pixies, and other spooky visitors. During the Great Thunderstorm of 1638, it is claimed that the devil himself paid a visit. Here, an abandoned old winch by a deserted quarry adds to the desolate atmosphere.

Fog brings rain when it rises, but clear weather when it falls.

Fog typically clings to the ground, but it can extend up to higher altitudes, depending on temperature. As the temperature rises, so does fog. Water vapor lying close to the ground has nowhere to fall, but at a higher altitude the same cloud can release precipitation—it is, in effect, a rain cloud and no longer fog. By the same token, a low cloud close to the ground—that is, fog—forms only under clear skies when surface temperatures are cool enough.

THE MISTS OF DARTMOOR

For lovers of mysteries and horror films, no setting conveys a sense of eeriness more than the swirling mists of a forest or moor—who knows who or what will emerge from the murk? There are many legends about the dangers that can lurk in the mist. In England the witch Vixana conjured up mists to confuse and waylay travelers. Dartmoor in Devon, England (now a national park), is famous for its mists. The mists of Dartmoor were believed to have inspired Arthur Conan Doyle's masterpiece *The Hound of the Baskervilles* (1901), which he wrote in a room at the Duchy Hotel looking out over the moor and the grim, fortresslike Dartmoor Prison in Devon. The hounds of the title were based on the legendary "wisht hounds" (black dogs with baleful red eyes) that hunted down sinners on orders of the devil. This is how Conan Doyle described one of these terrifying creatures: "A hound it was, an enormous coal-black hound.... Fire burst from its open mouth, its eyes glowed with smoldering glare, its muzzle and hackles and dewlap were outlined in flickering flame." A terrible thing to see springing at you from out of a mist!

FOG AND MIST

The difference between fog and mist is simply one of density: Fog contains more water droplets and so it is denser than mist. Fog usually forms when the moisture in the ground begins to evaporate. As it does, it is cooled by the air and condenses into droplets—the dew point—forming fog.

The amount of moisture that air can hold depends on temperature. If the air is cooler, it is able to hold less moisture.

When air is heated by the sun, it acquires a greater capacity to hold more moisture, promoting evaporation. In that sense the sun "burns off" the fog. The process is reversed at night when the ground loses, or "radiates," heat.

Morning fog in May and September is followed by warm, sunny afternoons. Morning fog in November to February is followed by cold, overcast afternoons (or the fog persists throughout the day).

The altitude of a fog depends primarily on air temperature. Fog tend to lift as temperatures rise, but form again when temperatures fall back down, usually after sunset. In May and September, the sunlight will lift the fog or clear it altogether as the water vapor evaporates. Under those circumstances, afternoons will usually be clear. In winter, on the other hand, fog tends to linger because temperatures are generally much lower and the difference between morning and afternoon temperatures is comparatively small. Even though winter sunlight may be bright, it does not carry much warmth, due to the angle of the sun, so it will not have the energy to lift the fog.

Fog from seaward, fair weather; fog from land brings rain.

This saying has some validity under most circumstances, given the fact that most fog forms under calm, clear skies, provided the ground is cool enough. Sometimes, though, fog can be treacherous. This is especially true in the North Atlantic between the U.S. coastline and the Gulf Stream (a strong current of warm water) where dense fog is common. It was in just such a fog that two ocean liners—the *Andrea Doria* and the *Stockholm*—collided, with considerable loss of life, in 1956.

Heavy fog in winter, when it hangs below the trees, is followed by rain.

Fog in winter is formed under different conditions than in summer; these clouds have collected water after moving over a body of water before passing over land. Sometimes, though, fog will be trapped by a high-pressure system overhead, creating an inversion. Heavy fog can combine with pollutants to form smog, a dangerous brew that has been dubbed "poison fog." This happened in 1930 in Liège, an industrial city in Belgium, killing 64 people and poisoning hundreds in only a few days. London has endured some famously thick fog. One chronicler, John Evelyn (1620–1706), wrote in 1661 of a fog so thick "that people lost their way in the streets, it being so intense, that no light of candles or torches yielded any or but very little direction." In December 1952 the "Great Smog" engulfed London for four days. At one point, the air was so thick with pollution that visibility was down to one foot. It is estimated that 12,000 people died as a result. In October 1948, also because of an inversion, the Monongahela River Valley in Pennsylvania was enshrouded by smog, which trapped sulfurous pollutants from steel plants for several days. As more people sickened and died, the factories shut down, but it finally took a good rain to lift the fog and clean the air—just as the saying insists.

Three foggy mornings will bring a rain three times harder than usual.

This saying comes from an old New Jersey almanac. Although there is no reason to think that three foggy mornings—or four or five, for that matter—will necessarily bring heavy rainfalls, so much moisture in the air would indicate the likelihood of some rain. Some fog clouds will lift from the surface under certain conditions and turn into rain clouds.

Foggy Conditions

For airline pilots, when visibility is reduced to 1,094 yards (1 km), it is classified as fog, but for earthbound motorists, 656 feet (200 m) is the usual yardstick.

TYPES OF FOG

Not all fogs are alike:

- Radiation or ground fog forms on clear, calm nights when the ground loses heat by radiation, cooling. Radiation fog adheres close to low ground, but as the night goes on, it can become more widespread and denser.
- Advection fog forms when very mild moist air moves over cold ground, a situation that is common in spring in the Northern Hemisphere. As southwesterly winds (that bring moist air) move over ground still covered with snow or ice, the lower layers of air cool rapidly, permitting fog to develop.
- Hill fog forms on upper slopes. As mild, moist air moves uphill, it cools and becomes so saturated that a cloud forms below the peak.
- Coastal or sea fog forms, often in spring and summer, when moist air cools enough to become saturated as it moves over cooler waters. Wind may blow the fog along the coast.
- Steam fog develops after a rainstorm when the ground is warm. The water from the rain will begin to evaporate from the ground, but if the air above it is no longer capable of holding so much moisture, it condenses. The result is the steamlike effect that gives the fog its name.

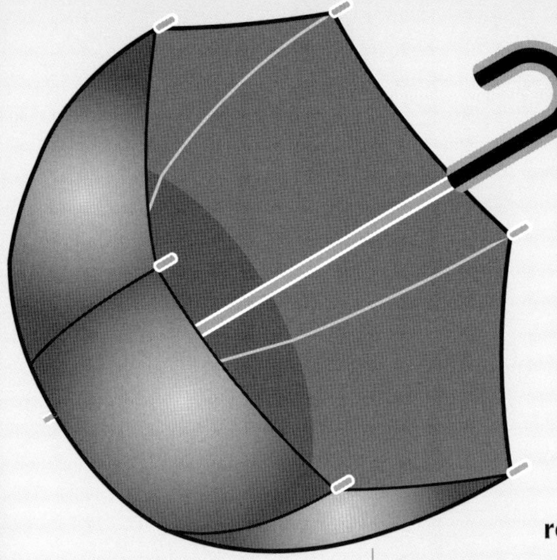

RAIN

Rain goes away and comes again another day, in the words of the nursery rhyme. It has been an enduring preoccupation of humans for millennia, accounting for far more sayings than snow, no doubt because all regions of the world experience rain. It seems that almost every plant, animal, and insect has been recruited to forecast rain.

Umbrellas

Although commonly associated with rain, umbrellas were originally used for protection against the sun. The umbrella has a venerable history extending back about 4,000 years. Archaeologists have found evidence for umbrellas in Egypt, Assyria, Greece, and China. The Chinese were the first people to waterproof their umbrellas for use in rain. The author of a 1720 English dictionary commented that the umbrella was "now commonly used by women to shelter them from rain."

Mackerel sky,
mackerel sky,
not long wet,
not long dry.

It is the herringbone pattern formed by long, wispy cirrus clouds that produces the mackerel sky referred to in this saying. These cirrus clouds—called mares' tails because of the resemblance—form at high altitudes and indicate that a low-pressure system carrying moisture-laden clouds is moving in from the west. This usually means an increase in wind speeds, a shift to blustery easterly winds, and increased cloud cover and precipitation.

Cats and dogs eat grass before a rain.

There are many sayings that rely on the habits of cats and dogs to forecast the weather. Cats and dogs may eat grass, but it has nothing to do with the weather. Some researchers believe that dogs eat grass as an emetic when they feel ill.

If a cat sneezes once, it will rain.

Sailors traditionally believed that cats possessed magic in their tails, so it was best to keep them content. A preponderance of weather lore about cats focused on the animals' ability to forecast inclement weather—if a cat licked its fur against the grain, for instance, it meant hail; if its eye grew wide, it meant it would rain. There is, however, no truth to most of this lore.

Flies bite more before rain.

Moist weather makes flying for insects more difficult; so they may be more prone to hovering closer to the ground—at human level. Under humid conditions, humans make more delectable targets, too: The heat causes them to perspire more, and as the atmospheric pressure diminishes, body odors tend to become more prominent. So flies and other insects might indeed prove more annoying before a rainfall.

PROBABILITY OF RAIN

Meteorologists use what are known as probability forecasts to estimate the likelihood that a given area will get measurable precipitation during the forecast period—a 25 percent or 40 percent probability, for example. To arrive at the forecast, a meteorologist needs to assess current and past weather patterns and then try to assess how these patterns are likely to change in the near future. Once it is determined that precipitation is likely to occur, the forecaster has to estimate the form it will take: Showers, for example, are usually localized, whereas a steady rainfall is usually more widespread. The basic procedures for making a probability forecast are little different from those needed to produce a regular forecast, but probability forecasts require more time and checking. In any case, they do not replace regular weather forecasts, but rather supplement them.

The accuracy of probability forecasts is limited since they cannot be used to predict when, where, or how much precipitation will occur. Saying that there is a 50 percent chance of rain tomorrow does not mean that it will rain half of the day, nor does it tell you how much it will rain or exactly where (if it does rain at all). Statistically speaking, the accuracy of probability forecasts can be assessed only by calculating the reliability of hundreds, if not thousands, of forecasts. If a forecast predicted a 30 percent chance of rain 100 times, and if it rained on 30 of those occasions, then you would say that the probability forecast was 100 percent reliable.

Probability forecasts are most reliable over the course of a 12- or 24-hour period, but are much less accurate for any given interval of time. For instance, the British Met Office's 24-hour forecast of rain boasts 83 percent accuracy for a precipitation forecast for the next day. However, a 1996 study found that the probability of rain on an hourly timescale was around 0.08 percent. In other words, forecast accuracies of 83 percent for a 12-hour period did not translate into a similar percentage if you tried to rely on it for 2:30 or 5:00 P.M. In fact, the author of the study concluded that the Met Office's forecasts of rain were over twice as likely to be wrong as right.

If the rooster crows at night, be is trying to say rain is in sight.

Roosters feel the changes in pressure, temperature, and humidity in the air, just as humans do—probably more so because air is trapped in their feathers. The air will exert less pressure on their bodies because of falling atmospheric pressure. These changes may make roosters restless or uncomfortable, which might prompt them to crow at night, hours before they ordinarily would. Some traditional farmers may still use roosters to anticipate rain, but few meteorologists are likely to do so.

Rain Rain Go Away!

The world record for the greatest amount of rain recorded in 24 hours was in the island of Réunion in the Indian Ocean, when 72 inches (182.5 cm) fell on January 7 and 8, 1966.

The crow with loud cries a sudden shower foretells. In single file they fly up and over the bills.

Although crows tend to be social when they feed, they roost and seek shelter on their own. They are known to fly off in single file when a storm approaches, which is why this saying has some currency in several southern states in the United States and in some parts of England.

When the cow scratches her ear, it means a shower is near; but when she thumps her ribs with her tail, expect lightning, thunder, and hail.

This saying is based on the observation that the hairs inside a cow's ear act as a barometer, responding to the changes in atmospheric pressure that can portend rain, so it is possible that the cow scratches in response. Static electrical discharges that accumulate in the air prior to a violent thunderstorm may cause the cow's hair to stick up. According to some farmers, that accounts for the movement of its tail. Little scientific study has been carried out to corroborate these reports.

When the ditch offends the nose, look for rain and stormy blows.

There is some scientific basis—aside from the evidence provided by your nose—to support this saying. Bogs, drains, and ditches typically contain stagnant water that give off an unmistakable and unpleasant odor. The odor is produced by decaying organic debris—leaves and grass—that emits methane and other gases. Under high-pressure systems, which are usually associated with fair weather, odors found in stagnant water are trapped in the mud. However, when atmospheric pressure diminishes when a storm approaches, bubbles of these gases expand, rise to the surface, and then break loose, making the surrounding air much more pungent.

When smoke descends, good weather ends.

Under clear skies and calm winds, smoke will continue to rise. However, when the atmospheric pressure lowers and clouds roll in, portending rainfall, smoke no longer rises as quickly or as high. The clouds, in effect, acts as insulator. Winds that arrive in advance of a storm system may also cause the smoke to curl downward.

When chickens scratch together, there is sure to be foul weather.

The belief that chickens' behavior changes before a storm is due to the fact that their feathers trap air, so that they are more sensitive to lower atmospheric pressure, which indicates an approaching storm.

Atacama Desert, Chile
The Luna Valley, in the Atacama Desert near Arica, is one of the driest places on Earth. Its name—Valley of the Moon—was reputedly coined by astronaut Neil Armstrong during a visit there.

RAIN FACTS

- Cherrapunji, India, has the greatest average yearly rainfall in the world, weighing in with a yearly average of 450 inches (1,143 cm), measured over 74 years.
- Based on a 59-year average, Arica in Chile, receives only 0.03 inches (2.032 cm) of rain every year, and did not receive any rainfall for 14 years in a row.
- Bagdad, California, holds the U.S. record for the longest period with no measurable rain—767 days, from October 3, 1912, to November 8, 1914.
- In January 2005, California was deluged by 10 inches (25.4 cm) of rain in a few days, as a result of a subtropical jet stream—dubbed the "pineapple express"—while the nearby Sierra Nevada were buried under about a foot (30 cm) of snow.

THE RAIN OF MUD

Rain can assume a variety of forms—and colors. Take, for instance, the phenomenon of blood rain. The Roman historian Livy (59 B.C.–A.D. 17) and the Greek writer Plutarch (c. A.D. 45–125) both wrote of blood rain falling on Rome on different occasions—never a good sign. St. Gregory of Tours noted that "real blood from a cloud" fell over Paris around A.D. 582. There have been other sightings, but most of those can probably be attributed to large accumulations of red dust, possibly gathered by the wind from deserts, lifted into the atmosphere, and then deposited far away.

There are also reports of yellow rain, stirring alarm among people who believed it to be sulfurous in nature. Pollen from flowers and pine trees seems to be the more likely explanation, though.

In March 1879, mud fell for an hour in parts of Nebraska. It is likely that the mud resulted from a combination of dust and water in the air. Iceland experienced a related phenomenon in 1755 when a black rain fell, an event probably linked to a nearby volcanic eruption. Even when there is a rational cause, the impact of colored rain on those who see it cannot be underestimated.

When windows won't open, and the salt clogs the shaker, the weather will favor the umbrella maker!

Windows with wood frames tend to stick when the air is full of moisture. The moisture swells the wood, making windows and doors more difficult to budge. By the same token, salt is very effective at absorbing moisture, so it clumps together rather than pouring out. As moisture collects in the air, there is greater likelihood of precipitation.

The odor of flowers is more apparent just before a shower than at any other time.

This saying implies that the accumulation of moisture in the air before a storm somehow intensifies a flower's scent. Floral scents have two major purposes: to attract pollinators and to repel harmful insects. (It might be argued that they also attract humans, who will lavish great attention on them—certainly an evolutionary advantage.) In most flowering plants the petals act as a miniature factory, converting oils into scents. How many oils are used depend on the type of plant—a modest 7 to 10 in snapdragons, or over 100 chemicals in orchids. The oils are volatile compounds because they evaporate easily in warm weather. Floral scent is related to several factors, including the environmental conditions during the flowering period. Sweet peas, for example, are not as fragrant in hot weather as they are in cool weather. In that respect humidity will only be one and not necessarily the most important factor in the strength of a floral scent.

Corn fodder dry and crisp indicates fair weather; but damp and limp, rain.

Corn is sensitive to weather, which for long stretches during the growing season it may be either too wet or too dry for optimum corn growth. Under certain circumstances, even average seasonal weather will produce stress in corn. Stress can retard growth until its source is removed. Although stress in corn is not hard to detect, a casual observer is not likely to notice any symptoms. Corn stress is often due to a lack of water, especially in the "corn belt" in the U.S. Midwest, where dry conditions are common. However, excess moisture can also create stress, so very humid weather would manifest itself in the look of the plant's leaves. Stress can arise from many other factors—adverse soil moisture and temperature conditions in combination with nutrient deficiencies, diseases, insects, and weeds. There has to be an adequate reserve of moisture in the soil to make up for the moisture that will be lost to the atmosphere as a result of evaporation. Evaporation is at its highest level on windy, hot, sunny days with low humidity; so if a corn crop is to avoid stress, the available soil moisture must be more than enough to compensate. Under cloudy skies, high humidity, and cooler temperatures, atmospheric evaporative demand is low, which means that less moisture in the soil is necessary.

If toadstools spring up in the night in dry weather, it indicates rain.

Mushrooms proliferate when the weather is moist; in fact, without moisture, mushrooms could not grow at all. Mushrooms need water in the atmosphere because they have no roots, which deprives them of a way to reach sources of water in the soil. The saying, then, cannot be taken seriously; mushrooms could not sprout up overnight in the absence of water, and since they are unable to get it from the soil, they would have to find it in the atmosphere.

Toadstools
Toadstools, more commonly known as mushrooms, need water—a lot of it. In addition, mushrooms usually need darkness or shade; only a few species are able to tolerate sun for more than a few hours at a time.

When clouds appear like rocks and towers, the earth will be washed by frequent showers.

The clouds in this saying are large towerlike cumulonimbus clouds in which violent thunderstorms develop. These clouds are composed of water droplets at lower elevations and ice crystals at higher elevations. When clouds are no longer able to contain water (in whatever form) that has built up, precipitation results. Thunderstorms are also characterized by intense winds when updrafts of hot air rising off Earth's surface and downdrafts of cooler air from the higher reaches of the storm account for their turbulence.

Towering Clouds

Convective clouds such as cumulonimbus show great vertical development. These clouds, which are typically brilliantly white on top from sunlight, and dark on the bottom, can assume the form of towers, castles and mountain peaks. They tend to form on hot summer afternoons and can produce severe thunderstorms with lightning and heavy rain. Convective clouds are basically towering cumulus clouds. They are also called cumuliform from the Latin cumulus, *meaning "piled up."*

When a halo rings the moon or sun, rain's approaching on the run.

A halo appears around the moon or the sun when ice crystals at high altitudes refract the moonlight (or sunlight). That is a good indication that moisture is descending to lower altitudes, where it is likely to take the form of precipitation. It signals the arrival of a low-pressure system. The brighter the halo, the more probable a storm becomes. A halo is a more reliable indicator of storms in warmer months than during winter months. If a storm does develop, it is usually within 12 to 18 hours of the halo's appearance. Halos often turn into what is known as "milk sky." A milk sky may appear clear, but the sky's characteristic blue color is either bleached or barely noticeable. This is an indication of high, thick cirrostratus clouds that are harbingers of an approaching low-pressure system.

If clouds move against the wind, rain will follow.

Under some exceptional circumstances, this saying has validity. To assess whether conditions are improving or not, you can stand with your back against the wind and study the movement of the clouds. If the upper-level clouds are moving from the right, then it is likely that a low-pressure area (with rain-bearing clouds) has passed and weather will improve. If the opposite is the case, and clouds are moving from the left, then weather is more likely to deteriorate. This version of the "crossed winds" rule, as it is called, applies only to the Northern Hemisphere;

MONSOONS

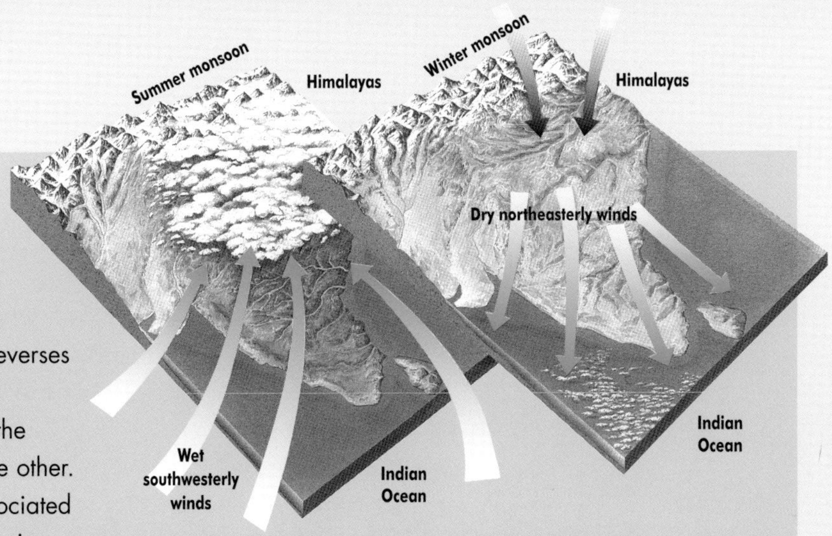

A monsoon is any climatic system in which the moisture increases dramatically in the warm season. It is characterised by a wind pattern that reverses direction with the seasons—blowing from the southwest during one half of the year, and from the northeast during the other. Although monsoons are frequently associated with seasonal winds that bring heavy rain to the Indian Ocean and Arabian Sea, they occur elsewhere in the world, including Australia and the United States. The term is derived from the Arabic Mausin, or "the season of winds."

Monsoons require certain conditions to occur, among them a wet summer, a dry winter, a hot land mass, and a cooler ocean. In India, for instance, the land heats up faster than the surrounding Indian Ocean does. Monsoons develop when heat begins to rise off the land in summer, which creates an area of low pressure. It is this low-pressure system that causes winds to blow landward from the Indian Ocean (a sea breeze). The winds carry moisture from the water. This warm, moist air is diverted by the Himalayas, propelling the air upwards, where it cools, condenses and forms rain clouds. Without monsoons the region's agricultural economy would suffer, but downpours can bring devastating floods, especially to Bangladesh whose land mass is mostly at sea level. The monsoon pattern reverses in the winter as the land cools and winds blow out to sea.

The monthly total of rainfall averages 8 to 12 inches (20 to 30 cm) over India, with the largest precipitation occurring during the middle of the monsoon season in July and August.

The North American monsoon—also called the summer, desert, Southwest, Mexican, or Arizona monsoon—occurs from late May or early June into September; it originates over Mexico and spreads into the southwest United States by mid-July.

the reverse holds in the Southern Hemisphere. Clouds moving parallel to but against the wind may indicate an approaching thunderstorm. Winds blowing out of a storm typically travel opposite to the updraft zone of the storm, while clouds carried by winds at higher altitudes will appear to be moving against the wind at ground level. It is not uncommon for winds at the surface to flow in the opposite direction of upper-level winds at frontal boundaries or to the north of the frontal zones. These contradictory wind patterns do not, however, signal any change in weather; instead they signal that whatever the weather, fair or rainy, the conditions will persist for at least several hours.

Rain before seven, clear by eleven.

The temperature of the air is likely to be cooler at night or early in the morning than later in the day; if temperatures are cool and the air above the ground becomes supersaturated—meaning that it cannot hold any additional moisture—the water vapor will condense and form rain. As the heat of the new day begins to warm Earth's surface, much of the moisture will evaporate, and so the likelihood of precipitation will decrease. Late-night and early-morning rains may also represent the last precipitation of a passing weather front. But the departure of one front does not necessarily mean that another front will not move in, bringing more rain later in the day. So this saying has more credibility when it refers to nonfrontal weather. If ground temperatures are sufficiently high in late afternoon or early evening, rain showers can develop and continue until early morning before dissipating.

Hear the whistle of a train?
'Tis a sign it's going to rain.

This saying, dating from the early part of the twentieth century, is relatively recent by folklore standards. The mournful sound of a train whistle heard far away

PONDS AND MARSHES

Ponds are made up of still water; because they are usually much warmer than rivers and streams, they are capable of supporting a variety of plant and animal life. Ponds with colder waters will not be able to support flora and fauna to the same extent, however. In areas with a great deal of vegetation, ponds will usually see the formation of scum—the dead and decaying organic matter that condenses on the surface of the water. Algae helps this process. The presence of the decaying vegetative matter provides a nutrient-rich environment. Ponds are formed by digging a hollow or by trapping water in a valley with a dam.

When ponds become filled with sediment, they may turn into freshwater marshland. Marshland consists of treeless land in which the water table is at, above, or just below the surface of the ground. Saltwater marshlands also exist and can be found on coastal tidal flats. Inland salt marshes occupy the edges of saline lakes. The health of marshland is dependent on the environment for its nutrients, water circulation, and the type and deposition of sediment. Weather also plays an important role, especially in the prairie country in North America, where the marshland undergoes a periodic cycle of renewal that is instigated by drought.

from the tracks would often let farmers know that they should be prepared for heavy, prolonged rain. The key lay in the clarity of the sound. Under ordinary conditions, sound is muffled or distorted by dust and pollutants in the air. However, when the sky is covered by low, thick clouds—clouds heavy with moisture—the acoustical effect is very different: The sound, rather than dissipating, is amplified. It is also possible that the increased velocity of the wind pushed ahead by the storm may also account for the effect.

Seagulls

Turbulent conditions in the atmosphere make it more difficult for sea birds to fly or hunt for fish. Seeing seagulls huddled together on the sand, though, does not mean that the weather will turn bad—in all likelihood it already has.

Seagull, seagull, stay on the sand;
it's never fair weather if you're over land.

When coastal conditions are tranquil, seagulls obtain their food from the water and shoreline. When winds pick up to an unusual degree, however, seagulls are forced to find food sources farther inland. Turbulent gusts make it harder for them to fly and also indicate that the water will be choppy, too, making it more difficult to pluck fish out.

St. Swithin's Day, if it does rain
full forty days, it will remain.
St. Swithin's Day, if it be fair
for forty days, 'twill rain no more.

According to English tradition, St. Swithin's Day, July 15, is supposed to forecast rain over the next month and a half. The day is named for the Saxon Bishop of Winchester; as he lay on his deathbed, he asked to be buried outside so that "sweet rain" would fall on his grave. Years later, though, on July 15, 971, the monks transferred his remains to a shrine inside the cathedral. A rainstorm was said to have taken place on that day and continued for weeks afterwards as a sign of the bishop's displeasure at having been relocated. In the folklore that developed, a rainy St. Swithin's Day would lead to 40 days of rain, while a pleasant St. Swithin's Day would lead to 40 days of good weather. A study by Britain's Met Office found that on 55 occasions when rain fell on St. Swithin's Day, the prediction failed to materialize. St. Swithin is not the only rain-forecasting day: In France the feast day of St. Medard is supposed to forecast weather, and in Russia, the Feast of the Protecting Veil of the Mother of God serves the same meteorological function.

If birds fly low,
expect rain and a blow.

Birds are more likely to fly at a higher altitude when a high-pressure system prevails and winds are calm. A low-pressure system, on the other hand, presages stormier weather with blustery winds and air that is less dense, making it more difficult for birds to fly.

Much foam on a river foretells a storm.

This Scottish saying is not a good prognosticator, since foamy white water is formed due to conditions in the river itself. Foamy water occurs when a river's gradient drops enough to give rise to a bubbly, or aerated and unstable, current. The frothy water appears white. The term is also used loosely to refer to less turbulent, but still agitated, flows.

Hark how the chairs and tables crack!
Old Betty's nerves are on the rack....
'Twill surely rain; I see with sorrow.
Our jaunt must be put off tomorrow.

This saying reflects the belief that it was possible to "hear weather." It is true that atmospheric changes can occasionally influence acoustics. An approaching storm, for example, causes dust, pollutants, and other sources of acoustic distortion to dissipate. This makes it possible to hear sound with unusual clarity.

Trout jump high
when a rain is nigh.

Several events have to happen before trout start jumping; but when they do so, there is a reasonable chance that rain is in the offing. Under conditions of low pressure, which signal the approach of a storm, methane and other gases from decaying organic matter otherwise trapped in the bottom of lakes or ponds will be released. When the diminishment in atmospheric pressure triggers the release of these gases, microscopic organisms that make their home in the plant debris will be dispersed. For small fish, this is an invitation to an unexpected dinner. As the

RAIN GAUGES

The common rain gauge in use today was only invented a century ago. It consists of a large cylinder—19.6 inches (50 cm) high and 7.8 inches (20 cm) in diameter—with a funnel and inside it a smaller measuring tube. The funnel directs the precipitation into the measuring tube. It is divided, marked, and labeled in smaller parts (increments) in tenths and hundredths, allowing meteorologists to make precise measurements—to 0.01 inch (0.254 mm). When water reaches the 30 marker, for instance, you would say that you had 30 one-hundredths of an inch of rain, or 0.30 inch (7.62 mm). When rain is less than 0.01 inch (0.254 mm), meteorologists refer to the rainfall as a "trace" of rain.

Rain gauge amounts are read either manually or by sensors at an automated weather station. In some countries professional meteorologists are supplemented by a network of volunteers who collect precipitation data in sparsely populated areas. The larger the gauge, the more accurate the readings. That is because a larger gauge is capable of taking a more representative sampling of the precipitation. Larger gauges have an additional advantage in that they are better at collecting drizzle and snow.

Rain Gauge

The standard rain gauge, little changed since this version in 1922, can measure up to about 2 inches (5 cm). In the event of a particularly heavy rainfall, water will overflow the measuring tube, spilling into the cylinder containing it. (This is why the cylinder in which it rests is also known as the overflow tube.)

small fish race to feed on the tiny organisms, their sudden activity may spur larger fish such as trout to take an interest in them. Their frenetic activity in response may sometimes, but not always, be viewed as a kind of barometric pressure reading.

When the glass spills over, so will the clouds in a little while.

This saying, which comes from Cape Cod, Massachusetts, refers to a version of a barometer that Cape Codders called a weatherglass. Early Americans did not make very much use of barometers, although they had become popular in England. America made its own contribution to barometric technology in the 1800s with the introduction of the weatherglass—a sealed glass with a spout containing water. The water registered changes in atmospheric pressure by rising or falling. If the pressure built up sufficiently on the outside of the glass, it would pop the seal at the end of the spout, causing the water to spill over. That would mean that rain was likely. If a high-pressure system was moving into the area, then the pressure in and outside the glass would equalize, forecasting drier conditions.

If the ash before the oak,
we shall surely get a soak.
If the oak before the ash,
we shall only get a splash.

This old English rhyme refers to the time when the oak and ash tree first show their buds in early spring. Its value as a forecaster is limited by the fact that oaks almost always bud about two weeks before ash, so that would almost guarantee that rainfall would be modest for the remainder of spring.

Sharp horns on the moon threaten bad weather.

The moon in this instance is supposed to predict precipitation because it is perceived as being in the shape of a bowl, which means that it is filling with water or snow. If its "horns" are tipped to the side, some people believe that precipitation will descend.

Rain long foretold, long last.
Short notice, soon past.

This weather proverb may be based on the observation that storm systems that will bring heavy rainfalls announce themselves in advance: The atmospheric pressure diminishes, winds pick up, a variety of clouds fill the sky, and the temperature may drop. A passing shower, however, often gives little or no warning at all. The sky will darken suddenly, the rain will come down, and then 20 minutes or an hour later the rain will be gone and the skies will clear again.

To talk of the weather, it's nothing but folly;
for when it's rain on the hill, it may be sun in the valley.

The usefulness or futility of talking about the weather may be a subject of debate, but the truth of this weather proverb is not. Weather reports often speak of scattered showers, especially in summer. Driving rainstorms will typically cover the entire sky for hundreds of square miles (or kilometers), but shower activity tends to be localized. A sun shower—where the sun can be shining while rain is coming down—is a good example of the phenomenon.

RAINBOW LEGENDS

Around the world, many legends have sprung up that center on rainbows. The most famous rainbow is the one that appears in the story of Noah.

In ancient Greece, rainbows were the sole responsibility of Iris, wife of the god Zephyrus. As a messenger between the gods and mortals, the deity shuttled back and forth dressed in dazzling multicolored robes. It is those shimmering robes gave us the word "iridescence."

In some North Carolina folklore, rainbows were considered a sign of impending calamity. One old tradition holds that a house overarched by a rainbow would experience a disaster in the near future.

RAINBOW FORMATION

Light bends when it travels from one medium to another—from air to water, for example—and as it does so, it changes direction. The ability of light to do all this is called "refraction." What appears to be white light is actually composed of many colors—that is, the spectrum. These colors have different frequencies, and they travel at different speeds in different mediums. A simple prism can separate out the colors because of its slanted surfaces. Light entering a prism will bend, and when it exits, it bends again. Since colors are traveling through the prism at different speeds, they will emerge one at a time in a process called dispersion. In this case the prism acts as the dispersing element. Something similar happens when the rays of the sun hit droplets of water in the air. Drops of water have the same effect that a prism has: They act to disperse light, which then separates out into colors or wavelengths. (Blues, greens, and violets have shorter wavelengths; reds and oranges have longer wavelengths. Traditionally there are seven colors in a rainbow, but in reality, it comprises a whole continuum of colors.) That is how a rainbow is formed. A rainbow is simply raindrops bending sunlight.

If short showers come during dry weather, they are said to "harden the drought" and indicate no change.

This saying from Scotland is an acknowledgement of a bitter reality. In lands that are starved for water—the sub-Saharan countries, for example—rains are seasonal and tend to persist for weeks or months at a time. A long spell of dry weather at those times is cause for alarm. A brief shower under such conditions would not do anything to relieve the drought or replenish the water supply. On the contrary, the soil would be too dry to absorb the sudden spurt of moisture. It would be far preferable to have several days of steady rain.

Rainbows

Many legends are associated with rainbows. In the Book of Genesis the appearance of a rainbow was a sign of God's promise to Noah that the world would be spared any future catastrophic floods.

If the rain falls on the dew, it will rain all day.

Generally speaking, dew forms in spring and summer as a result of cooling temperatures during the night, when air close to the ground cannot absorb water vapor. At the right temperature—the dew point—water condenses and forms dew. Warming temperatures after sunrise usually cause the dew to evaporate. But if it is cloudy, sunlight will not reach the dew, and with cooling temperatures near the ground, precipitation is more likely for several hours more.

Cloud Types
There are three major classes of clouds—high, medium, and low—which are based on their altitude. Clouds have many other attributes that make it possible to categorize them including their shape, coloring, transparency, or opaqueness, and associated weather systems.

CLOUDS

Clouds have acquired any number of nicknames—mackerel sky, mares' tails, wool bags, and packet boys, among them—depending on their appearance. By changing form and location, clouds predictably dominate a great deal of weather folklore. In the Northern Hemisphere, a heavy cloud from the southwest, for example, will portend a storm; but so, too, will many clouds coming in from the northwest.

The more cloud types at dawn, the greater the chance of rain.

This saying holds true mainly in summer, when towering cumulonimbus clouds appear in the sky. These clouds are the ones that can bring heavy showers, violent thunderstorms, lightning, and even tornadoes. They are often accompanied by other types of clouds, and start life as harmless cumulus clouds.

The higher the clouds, the finer the weather.

Clouds are formed by moisture that condenses out of rising air currents. These currents are made up of hot air that is being lifted off the surface of Earth. The higher the air must rise before condensation begins, the drier it was to begin with. Rain clouds usually form in the afternoons in summer because they depend on rising pockets of hot air for fuel. As the atmosphere is heated up by solar radiation, hot air (which is lighter than cold air) will ascend into the atmosphere, where the water vapor will begin to cool and condense, eventually forming clouds. When those clouds become saturated, precipitation will result.

When clouds of the morn to the west fly away, you may conclude on a settled, fair day.

This saying is based on the observation that most storm systems come out of the west and move toward the east. If storm clouds are moving out of an area in the morning, the weather is likely to be fair—but their departure does not necessarily guarantee that another storm system won't materialize before the day is through.

If cirrus clouds dissolve and appear to vanish, it is an indication of fine weather.

Cirrus clouds are higher-level clouds that develop in filaments or patches. They owe their brilliant white luster to the ice crystals that compose them. They occur in flat sheets and typically appear in isolation with large patches of sky in between them. They can appear in a myriad number of shapes, depending on the wind. Cirrus clouds vary in thickness, but almost always permit the sunlight to get through. They are often propelled by westerly winds—not necessarily the same as winds at a lower altitude—but move slowly relative to the motion of lower-level clouds. Cirrus clouds tend to develop on days when the weather is fair with lighter winds at the surface. (Cirrostratus clouds generally follow the same pattern.)

While they form in good weather, cirrus clouds are indicators of an impending change in the weather. These changes usually involve approaching cold fronts, which could bring thunderstorms or advancing troughs of low pressure. Cirrus clouds normally precede cirrostratus clouds, which are harbingers of an approaching thunderstorm. Sometimes cirrus (and cirrostratus) clouds may hang on even after a change in weather has occurred. However, when cirrus clouds do dissipate, as the saying indicates, it is likely that the weather will remain settled for some time.

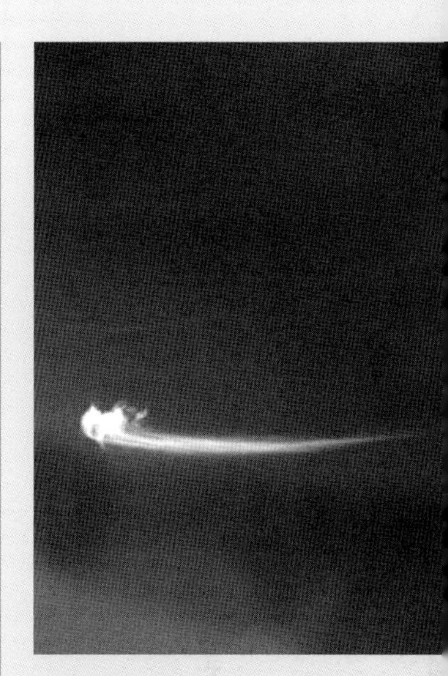

Cirrus Clouds

High-level cirrus clouds, like the cirrus uncinus, or "mares' tail," can signal the approach of stormy weather.

CLOUD CLASSIFICATION AND NAMING

Clouds are classified in a number of ways: by altitude, by height of cloud base, and by whether they are layered or convective. Layered clouds are named stratus clouds. Convective clouds are clouds that show great vertical development, such as cumulonimbus that bring thunderstorms, lightning, and considerable precipitation. Convective clouds are basically towering cumulus clouds (or cumuliform, from the Latin *cumulus*, meaning "piled up"). Clouds are also classified by the height of the cloud base (not the height of the top of the cloud). This classification system was first proposed in 1802 by the young English pharmacist Luke Howard. Howard was fascinated by clouds at a time when they were poorly understood and often referred to as "essences."

Cloud names are derived from the Latin words that describe them. Clouds whose names begin with the prefix "cirr" are located at high levels (cirrus clouds), while those with the prefix "alto" are mid-level clouds. (*Stratus* is the Latin for layer.) There are four "families" of clouds altogether: high, middle, low, and vertical (or convective). In addition, there are several types of clouds such as mammatus or roll clouds, which do not fit neatly into any category.

CLOUD TYPES AND ASSOCIATED WEATHER

Cloud Type	Description	Weather
High-level Clouds		
Cirrus	Thin and wispy, they can appear in a variety of shapes.	Approaching front and stormy weather.
Cirrocumulus	Small, rounded white puffs. Isolated or in long rows.	These winter clouds indicate fair but cold weather.
Cirrostratus	Sheetlike clouds covering the entire sky.	May signal precipitation within 15–25 hours.
Cirrus aviaticus	Wispy streaks.	If they linger, the air is relatively humid.
Mid-Level Clouds		
Altocumulus	Puffy clouds in bands or wavelike masses.	May signal precipitation within 15–25 hours.
Altostratus	A thin layer of grayish or blue-gray clouds.	Approaching storm. Produce snow and drizzle.
Low-Level Clouds		
Nimbostratus	Dark, low, uniformly gray clouds.	Widespread light to moderate precipitation.
Stratocumulus	Low, lumpy, puffy clouds, in patches or rounded masses.	Low, lumpy, puffy clouds, in patches or rounded masses.
Fractocumulus	Unstructured and disorganized puffs of cumulus clouds.	Produced as a result of a thunderstorm.
Convective Clouds		
Cumulus humilis	Small puffy clouds with minimal vertical development.	Fair weather. Never produce precipitation.
Cumulus mediocris	Puffy clouds with moderate vertical development and small protuberances and sprouts at their tops.	Associated with fair weather. Usually no precipitation.
Cumulus congestus	Resemble a cauliflower with a sharp outline.	Produce considerable precipitation, typically showers.
Cumulus castellanus	Narrow, very high towers. Tops formed of small puffs.	Suggest the onset of stormy weather.
Cumulonimbus	Heavy, dense clouds in the form of mountains or towers.	Heavy showers with thunder and lightning or hail.
Other Clouds		
Mammatus	Pouches that protrude from the bottom of a cumulonimbus cloud.	Indicate that a storm system is weakening.
Roll cloud	Low, tubelike shape. Seems to roll about a horizontal axis.	Associated with storm systems.
Shelf cloud	Usually curved and protruding from a cumulonimbus.	Indicates an advancing cold front.
Funnel cloud	Rotating column of air that extends from the base of a cloud (typically cumulonimbus or cumulus castellanus).	Very weak tornadoes with the potential to turn into full-blown tornadoes, often with supercell thunderstorms.
Wall cloud	Ragged, dark with signs of weak rotation.	Develops after an intense thunderstorm.

CLOUD TYPES

High-level cloud form where temperatures are very cold. As a result, water vapor freezes, forming clouds mainly composed of ice crystals. There are three basic types of high-level clouds: cirrus, cirrocumulus, and cirrostratus. Cirrus clouds, composed of ice crystals, allow more sunlight to penetrate and are associated with a warming effect. Cirrostratus clouds can build up to a thickness of several thousand feet (or meters) and blanket an entire sky.

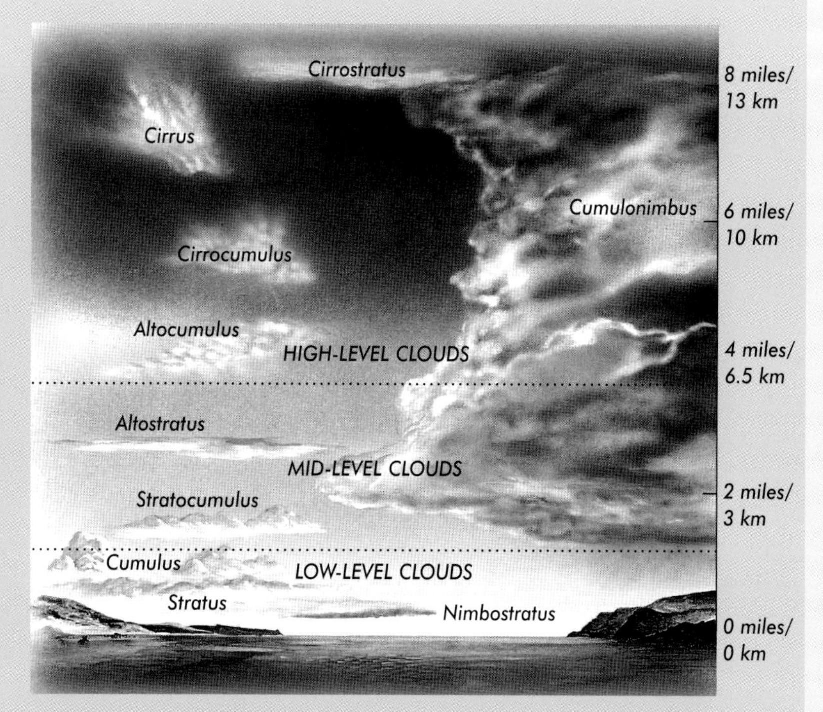

Mid-level clouds vary in their composition, depending on the season. In summer they are made up of both liquid water droplets and ice crystals. Mid-level cloud names are distinguished by the prefix "alto." There are two main types of mid-level clouds: altocumulus and altostratus. Altocumulus clouds are usually formed by a process known as convection, an updraft in an unstable layer of the atmosphere that culminates in thunderstorms. Altostratus clouds will form a thin layer that can cover the entire sky. These clouds—striated, fibrous, or in uniform sheets—are generally featureless.

Low-level clouds are generally composed of water droplets, but when temperatures are cold enough, these clouds may also contain ice particles and snow. There are two basic types of low-level clouds: nimbostratus and stratocumulus. Nimbostratus are sheets of dark, gray clouds. Stratocumulus clouds can form as rows, as patches, or as rounded masses; their size, thickness, and shape may vary considerably over a wide area.

Convective clouds also grow high into the atmosphere, which is why they are known as clouds of ascending air currents. They are seldom seen with other clouds. Because of their extension into the upper atmosphere, they can be composed of ice crystals, snowflakes, ice pellets, and supercooled water droplets, as well as regular water droplets in their lower levels. They are associated with unstable conditions characterized by rising air currents of warm air, and they can unleash torrential downpours and violent thunderstorms.

When the carry [current of clouds] gaes west,
gude weather is past.
When the carry gaes east,
gude weather comes neist.

This Scottish saying reflects the observation that bad weather tends to move from west to east. So if clouds are moving to the west, it would foretell a storm system moving in—the good weather is past—while if the system were moving in the opposite direction, pleasant weather would be in store, although the opposite may also hold true.

When clouds look like black smoke,
A wise man will put on his cloak.

This saying simply recognizes what seems obvious to anyone who has ever been caught unawares on a summer afternoon when the skies suddenly darken. Thick clouds laden with large droplets of water look darker than the fair-weather cumulus clouds. However, not all clouds that are low, thick, and dark produce precipitation. Before a drop or a snowflake can fall, it has to reach its "terminal velocity," which is defined as the fastest it can fall, given the pull of gravity and the resistance of the air holding it up in a cloud. Clouds are formed by rising air, so

CLOUD FORMATION

Clouds form from updrafts of warm air that are also known as thermals. The warm air contains water in the form of vapor or gas. Air can be lifted aloft under several circumstances: solar heating of the ground, the intrusion of a cold wedge of air (cold air is denser than warm air), the passage of warm air over a cooler surface (such as a lake or ocean), or the intervention of a mountain, if it is at an angle to the wind.

Clouds can also develop when two air masses—an arctic mass and a tropical maritime mass, for example—meet and mix. As the air continues to rise, it expands, loses energy, and cools. Meteorologists say that it is cooled below its saturation point (the dew point)—that point where the air can no longer hold a given amount of water vapor. The water vapor condenses into droplets that form clouds. At higher elevations, the water droplets become supercooled or turn into ice crystals.

Certain types of clouds such as cumulonimbus contain a mixture of water and ice crystals. The formation of a raindrop or snowflake requires

the only way that water droplets or ice crystals can fall is if they can descend at a faster rate than the air is rising. Under cooler temperatures in the cloud, the water vapor carried aloft on the air current from the ground begins to condense, forming water droplets. But the speed at which this process can happen presented a puzzle to scientists in the early part of the twentieth century: Rain clouds can develop in about half an hour, which isn't enough time for condensation to produce rain, given the fact that it takes up to a million liquid droplets to make one raindrop (see Coalescence on page 120).

Ice crystals, however, will grow quickly enough to start precipitating within such a short time. Crystals act as nuclei—seeds that water droplets are attracted to and cling to. But then scientists realized they were looking too low in the clouds for an explanation. Most thunderstorms, which typically occur in summer, begin as snowstorms in the upper atmosphere; it is just that by the time observers on Earth see it, the ice (snow) has melted into water because of higher temperatures it meets on its way down. To be sure, ice is not a necessary ingredient for rain to occur; in the tropics, for instance, water droplets keep bumping into one another and merge, so that they eventually grow large enough to reach their terminal velocity. But in general, many low clouds are simply unable to grow ice (because temperatures are not cold enough). That is why some dark clouds may look like they should yield rain, but do not. They do not have the capacity to grow their raindrops.

How Clouds Are Formed
Clouds form from rising air produced as the ground is warmed by the sun. The rising packets of air—thermals—are pushed farther up if they encounter hills or mountains. Colder, faster-moving air can also contribute to the process as it drives a wedge under the warmer air and lifts it into the atmosphere.

a seed or nucleus around which to form: for example, an ice crystal. The water in a typical cloud can weigh up to several million tons. The cloud can hold all this water (in whatever form) so long as air currents below it cooperate and the amount of water does not exceed the capacity of the cloud to keep it suspended. However, if the cloud becomes saturated, then the water will begin to fall; whether it falls as rain, hail, or snow, or some other type of precipitation will depend on the temperature in various layers of the atmosphere and on the ground.

COALESCENCE

Hundreds of millions of tiny water droplets form in clouds as a result of water vapor condensation carried aloft by relatively warm air currents. But for these droplets to develop into raindrops, the droplets have to merge—this process is known as coalescence.

In more stable environments the water droplets tend to remain apart because of air resistance, but clouds are always in a dynamic state for any number of reasons, for example, shifting winds or rising temperatures. Turbulence will stir up the droplets and cause them to collide and merge, forming ever-larger droplets that will overcome air resistance and so fall downward as rain.

The longer the dry weather has lasted, the less is rain likely to follow the cloudiness of cirrus.

The wispy, thin cirrus clouds, which are found very high in the sky, serve as indicators as to which way the wind is blowing. They are a sign of fair weather. Cirrus clouds produce no precipitation, so their appearance after an extended period of dry weather would offer no chance of relief any time soon, as the saying states.

Cirrostratus clouds typically precede wind and rain; the more cirrostratus clouds observed in the sky, the greater the likelihood that storms will occur within a short time.

Sheetlike cirrostratus clouds are formed at higher altitudes and owe their whitish, fibrous appearance to the ice crystals that compose them. Even though they can cover the entire sky and measure several thousand feet (meters) thick, they are relatively transparent. They may announce their presence as a halo around the sun or the moon because of the refraction of their ice crystals from sunlight or moonlight. Cirrostratus clouds tend to thicken as a warm front approaches, because of an increased production of ice crystals. They indicate precipitation within 12 hours.

No Beehive tonight...so prepare for unsettled weather.

The "beehive" of the saying refers to Praesepe, sometimes called the Beehive Star Cluster, one of the brightest galactic star clusters in the sky. It is situated just a couple of degrees to the right (or west) of Jupiter. Since ancient times the cluster has been used to predict weather. Because it looks like a misty patch of light to the naked eye, Greek writers called it the "Little Mist" or "Little Cloud." However, if the Little Cloud could no longer be seen, it was a sign that a storm was approaching. In that respect it served much the same purpose as a halo around the sun or the moon did. Meteorologists now know that prior to the arrival of unsettled weather, high, thin cirrus clouds (made up of ice crystals) show up in the sky. Although they are thin enough to permit a hazy glimpse of the moon, the sun, or even the brighter stars, they do conceal a more distant cluster like the Beehive.

HALOS

A great deal of weather lore is based on the association of halos and impending storms. Halos are produced by the reflection and refraction (or bending) of sunlight or moonlight off ice crystals in the upper layers of the atmosphere. (Rainbows are also caused by a similar process.) Put another way, ice crystals scatter sunlight or moonlight into different angles.

Halos would not be possible if not for the capacity of light to bend. Some light is reflected by the ice crystals—it bounces off—while other light penetrates the crystal. But as it does so, it is bent to a different direction or angle and emerges from a different side.

Halos can occur at either 22- or 46-degree angles from the sun. The clarity and prominence of the halo depends on the type and quality of the ice crystals. If the crystals (which are hexagonal, like snowflakes), are very large, they can produce bright and sometimes colored light patches on either side of the halo; these are commonly called sun dogs or mock suns, and are technically referred to as *parhelia*. These ice crystals are found in high-level cirrostratus clouds. These clouds are an indication that a warm front is approaching, accompanied by stormy weather. Usually rain or snow can be expected within 12 to 18 (or sometimes 24) hours after a halo is sighted around the sun or the moon.

Counting Stars

In weather folklore, there is a belief that by counting the number of stars contained within the halo, it is possible to predict how far away the rain or snow is, with each star representing approximately 24 hours.

Streaky clouds across the wind foreshadow rain.

This Scottish weather saying refers to cirrus clouds that owe their thin, streaky appearance to the action of the wind. (They are often called mares' tails for that reason.) While they may form in fair weather, they typically foretell rain if they begin to thicken and cover the sky.

Raindrops Keep Falling...

Not all raindrops are shaped alike nor are they all teardrop-shaped. Smaller raindrops (less than .08 of an inch [2 mm] in diameter) tend to be more or less spherical; as they grow larger they show signs of flattening with the largest drops (about 0.2 of an inch or 5 mm in diameter) taking on the form of parachutes. If the raindrops are larger than this, the drops become unstable and fragment.

The fish-shaped cloud, if pointing east and west, indicates rain; if north and south, finer weather.

The fish-shaped cloud of the saying is most likely a cirrocumulus cloud, a patch or layer of cloud made up of tiny individual cloudlets at high level. It typically has a regular dappled or rippled pattern that can resemble the scales on a fish—a "mackerel" sky that may mean that unsettled weather is on its way. If there is warming below it, water in the cloud will rise and sink; in that process some of the ice crystals will melt, causing gaps in the cloud. The saying reflects the observation that storm systems tend to move from west to east.

Cirrocumulus clouds, when accompanied by cirrostratus ones, are a sure indication of a coming storm.

This saying is correct, but only up to a point. Cirrocumulus is a higher-level cloud—typically found at 20,000 to 40,000 feet (6,000 to 12,000 m)—that is characterized by a brilliant white appearance due to the ice crystals that compose it. However, it is distinguished by many small turret-like formations that indicate vertical turbulence within the cloud. Ice crystals within the cloud will cause the supercooled water to freeze. Once the supercooled drops are frozen, the cirrocumulus cloud will be turned into cirrostratus. This process can also produce precipitation in the form of virga—precipitation that evaporates before it reaches the ground. There is rain virga and snow virga. Virga has been described as resembling "a torn drape or a curtain hanging from the cloud." Thus, cirrocumulus clouds tend to be relatively short-lived. Cirrostratus clouds are very thin; their name is derived from the Latin meaning "curl" and "layer," or "spread out." Because cirrostratus clouds develop when cool, dry air meets warm, moist air, they are often harbingers of rain or snow. They also produce a halo around the sun or the moon; the effect is caused by the refraction of light from the ice crystals in the cloud. While it is true that cirrocumulus clouds in combination with cirrostratus clouds may point to wet weather ahead, it is probably more accurate to say that cirrostratus clouds that have developed from cirrocumulus clouds do so.

When mountains and cliffs in the clouds appear, some sudden and violent showers are near.

This saying refers to detached clouds that show considerable vertical development into the higher reaches of the atmosphere; at their tops they can assume the shape of mounds, domes, towers, or mountains. These "towers" and "mountains" are cast into high relief by the sun, while their base is usually dark and almost completely flat. These clouds generally contain a mixture of elements—water droplets, supercooled water droplets, ice crystals, snowflakes, and ice pellets—because they grow from regions above freezing to levels well below freezing. These clouds form in unstable weather as warm air is forced upward and cooler air within the cloud travels downward. These alternating currents of air account for the violent thunderstorms and other types of weather that these clouds can produce.

Clouds that show such vertical development come in a variety of forms. Cumulus congestus, for instance, produce heavy precipitation in the form of showers. If they continue to grow, they turn into cumulus castellanus clouds, which resemble narrow, high towers. They can transform again into dense, dark cumulonimbus clouds with anvil tops that can extend up to the tropopause—the border between the troposphere and the stratosphere. They can bring heavy, intense downpours.

DOES WALKING IN THE RAIN KEEP YOU DRIER THAN RUNNING?

An old wives' tale insists that it's preferable to walk in a shower rather than run because running will only make you wetter. It sounds counterintuitive. But is there any wisdom in it? Several studies have been conducted to find the answer. The European Journal of Physics in 1987 ran a study by an Italian physicist who contended that if the distance was sufficiently short, a person who ran would get 10 percent less wet than someone who walked—not worth the trouble.

A study conducted recently by meteorologists at the National Climatic Data Center in North Carolina, however, determined that over a distance of 330 feet (100 m), running was a far more efficient way of keeping oneself drier—40 percent more—than walking was.

OKTA SCALE

The okta scale is used to measure how clear or overcast the sky is. For the purpose of measurement, the sky is considered a hemisphere called the celestial bowl. What fraction of this "bowl" is covered by clouds has a corresponding value calibrated in eighths.

A measurement of 5.7 would be rounded off to 6. If the cloud cover and blue sky are judged to be about equal, then the cover is considered to be scattered. If there are more clouds than blue sky, the cover is said to be broken.

Clear: Zero oktas.

Scattered Clouds: 1–4 oktas.

Broken clouds: 5 oktas.

Overcast: 8 oktas, even if breaks (up to ½ okta) are present.

The pocky cloud or heavy cumulus, looking like festoons of drapery, forebodes a storm.

This Scottish saying describes the transformation of cumulus clouds into more ominous cumulonimbus clouds. Cumulus clouds are the familiar low-level billowy puffs with flat bases that look a little like cauliflower. (*Pocky* means "baggy.") They are at least as tall as they are wide, and they form on sunny days when pockets of warm air rise from the ground. Under these conditions, cumulus clouds indicate fair weather. However, if they show more vertical development so that they "look like festoons of drapery," building up into the middle or high part of the atmosphere, they turn into tall, deep, and dark cumulonimbus clouds. These produce violent thunderstorms with lightning, heavy rain, damaging winds, or tornadoes—virtual factories of bad weather, fueled by rapidly rising and sinking air currents.

When scattered patches or streaks of nimbus come driving up from the southwest, they are called by the sailors "prophet clouds" and indicate wind.

Cumulonimbus clouds, easily identifiable by their dark coloring, are rain bearers. (*Nimbus* means "rainstorm" in Latin.) The precipitation that these low-level clouds yield will depend on the prevailing weather conditions; it can emerge as rain, snow, hail, and sleet. Nimbus is usually mixed with another cloud type to describe not only that the cloud is carrying precipitation, but also the cloud's height. But when they appear in patches or streaks, as the saying states, it would suggest that the storm has passed and the clouds have broken up. The direction the remnants of the cumulonimbus clouds are blowing from would indicate to sailors the weather they could expect in the future. Once a storm passes, a cold front has may move in, bringing the winds referred to in the saying.

There are few finer days in the year than when the morning breaks out through a disappearing stratus cloud.

Stratus clouds are low-level clouds that are flat and range from gray to white, with a flat, featureless aspect. These clouds are essentially fog that has relocated to a level slightly above the ground. They are formed either through the lifting of morning fog or when cold air moves at low altitudes over a region. If they produce any precipitation at all, it is generally as a mist. Fog forms under clear skies at night (when the earth is sufficiently cool for water vapor in the surrounding air to condense and turn into liquid droplets) and in calm conditions. In the morning it usually evaporates due to the heat of the sun. Under those conditions you can generally expect a fair day ahead.

When the clouds go up the hill, they'll send down water to turn a mill.

As clouds move uphill, they will cool due to the higher elevation. As they do so, the water vapor in the cloud will begin to condense and form precipitation; precipitation may be rain or snow, depending on the altitude. This saying from England dates back to a period during the eighteenth and nineteenth centuries when mills had to rely on water power to operate.

OCCLUDED AND STATIONARY FRONTS

Occluded fronts are a phenomenon seen in the final stages of a storm when colder air replaces relatively warmer air at the surface. The air in the warmer region is lifted off the ground. This happens in two ways. In one scenario, the air rushing in behind the front is colder than the air ahead of the front and overwhelms the cooler air in front of it. (Colder air moves more quickly than warmer air.) The effect is similar to that of a cold front, and the result is called a cold occlusion. In a warm occlusion, the air behind the front is warmer than the air ahead of it and forces the cool air at the surface to rise up and over the warmer air moving in.

Occluded fronts always have sharply defined boundaries between cold air, cooler air, and warm air. By contrast, stationary fronts result from a standoff between warm and cold fronts, when neither can advance. These fronts can last for up to a week or more. A boundary between them can form thousands of feet (or meters) above the ground. These fronts are typified by wet, cloudy weather for several days running.

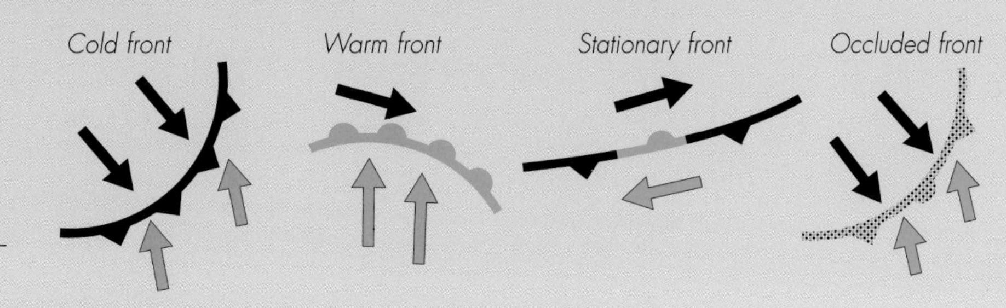

Fronts
Four types of fronts are represented on a typical weather map: warm, cold, stationary, and occluded. There are two types of occluded fronts—cold and warm.

Risk of Flooding

For coastal regions or for populated areas near rivers there is always a risk of flooding in the event of a major storm. However, as climate change brings warmer average temperatures and rising sea levels, the threat of catastrophic floods has become even greater. The devastation that Hurricane Katrina caused in New Orleans may offer a sobering example of what other vulnerable cities and low-lying areas may face in the near future.

FLOODS

Almost 5,000 years ago, some scientists believe, a large asteroid or comet crashed into the Indian Ocean, producing a tsunami about 13 times as big as the one that inundated Southeast Asia in December 2004. That almost every culture throughout history has had a flood myth, of which Noah and the Ark is the most famous in the Western world, demonstrates the fear and awe that floods inspire in human consciousness.

Two full moons in a calendar month bring on a flood.

This saying echoes another, somewhat less ominous one, which declares that if there are two moons—new or full—in the same month, the weather will be unfavorable and unsettled until the next new moon. Two full moons in a month is a relatively rare phenomenon, but it does happen because the cycle of the lunar phases is shorter than most calendar months, with the phase of the moon at the very beginning of the month usually recurring at the very end of the month. As a result, a month will have two full moons about once every 2.7 years, or 41 every century. The second full moon is commonly known as a "blue moon."

Although there is little evidence that two full moons in a given month can trigger a flood, the gravitational pull of the moon does have a significant influence over the creation of tides. Since the moon is much closer to Earth than the sun, it exercises proportionately more gravitational attraction. When the moon is closer to Earth, it can cause coastal waters to rise above normal levels. There are always two high-water areas on Earth at any given time: the area under the moon (because of gravity) and the area opposite the moon (because of centrifugal force). High tides move with the moon as it orbits Earth, producing two high tides and two low tides with every lunar day. A lunar day is slightly longer than one on Earth (24 hours and 50 minutes), so typically a coastal region will alternate between high and low tides about every six hours, but not exactly six hours. The highest tides in the world occur in the Bay of Fundy, a large, funnel-shaped inlet of the North Atlantic Ocean in Canada. There the tide can rise as much as 50 feet (15 meters) within a few hours.

The association with two moons and floods was incorporated into a flood myth of the Pima Indians of the American Southwest. "Then the roar of thunder and lightning enveloped the land. From the east the rains came, and for two moons it fell. All of the land except Crooked Top Mountain was covered with water. The Earth Maker spoke once again from the thunder clouds atop Kakatak Tamai. 'All good people will return to the desert valley to till the fertile soil, and all evil ones will be turned to stone,' he said. And so it was." The doomed humans were turned into rock faces, where they can still be seen, imploring the gods for mercy, in futility.

St. Mark's Square, Venice
Venice is deservedly famous for its canals, but in spite of valiant efforts the city continues to sink. Schemes to control flooding are being looked at; whether any of them can succeed, though, is unclear.

In April, Dove's flood is worth a king's good.

The Dove is a river in the English county of Devon; its name comes from the old English word *dubo*, meaning "dark." While flooding is often seen as a disaster, it actually can do some good, as this saying indicates. Wood deposited when the Dove overflows its banks creates niches for riparian vegetation to regenerate and thrive. Some 350 million years ago, the region where the Dove River flows was under a shallow tropical sea, with deep lagoons fringed by coral reefs. The sea creatures and corals that once flourished in that sea are memorialized in the fossilized limestone that forms much of what is now Dove Valley.

FLOOD FACTS

- Flash floods are responsible for more deaths than tornadoes or hurricanes.
- New Orleans is located 6 feet (1.8 m) below sea level; it was especially vulnerable when Hurricane Katrina struck in 2005 because it was surrounded by three large bodies of water: Lake Pontchartrain, the Mississippi River, and the Gulf of Mexico.
- The number of people in danger of potentially devastating floods may double—from 1 to 2 billion—by the year 2050, according to the United Nations University. Researchers blame climate change, deforestation, rising seas, and population growth for the elevated risk.
- In 2005 natural disasters associated with the flood season affected more than 210 million people in China, claiming 1,500 lives, and destroying over a million homes and at least 59,576 square miles (154,300 sq km) of farmland.
- The Johnstown Flood of 1889 in Johnstown, Pennsylvania, which was caused by the breach of a dam, was one of the worst in American history and claimed an estimated 2,300 lives.

FLOOD WARNING SYSTEMS

Flood warning systems provide advance notice to people vulnerable to impending flooding and provide advice on what precautions to take. These systems are found throughout the world in regions where flooding is common. Many of these systems offer "real-time" data on rainfall, storm water runoff, and general weather conditions.

In Australia, for example, the Bureau of Meteorology offers an alert watch in the event of immediate danger and a generalized flood warning if a flood is expected in a particular region. For towns along swollen rivers, the bureau will provide estimates as to when the river is expected to overflow its banks.

A river flood fishers' good.

Is the fishing really better after rains have swollen a river, as this Spanish saying suggests? As true with just about everything else regarding fishing, the answer is: It depends. Rain can have both positive and negative effects on fishing (if not on fish themselves): Heavy rains can create mud lines, and some anglers recommend fishing between areas of mud and areas of clear water, averring that the fishing is especially good in such zones. In addition, heavy rains can raise the water level, impelling fish to move higher in the water and so make it easier for fishermen to catch them. New water also tends to be muddy. That's good, some fishermen say, because the fish will not realize that anyone is there. That's bad, others say, because the water is too turgid to see where the fish are. Rain may also wash food into the river and helps oxygenate the water, factors that make fish more active—and more vulnerable to anglers. Other experts contend that fishing is actually better just before a cold front moves in or rain falls, but that by the second day into the front, the fishing becomes less productive.

Dog Days bright and clear indicate a happy year.

While we have come to think of the Dog Days as occurring during the summer—particularly late July and early August—this Welsh saying seems to apply to the rising of the Dog Star—Sirius—on or around January 1, when it reaches the meridian at midnight. In ancient Egypt, the Dog Star rose at the summer solstice and was regarded as a harbinger of the annual flooding of the Nile, without which the country could never have flourished or sustained its people. The name of the star is derived from the Greek *Seirios* ("glowing" or "scorching") by way of Latin. True to its name, it is among the brightest bodies in the sky; it became known as the Dog Star because it is a major star in the Big Dog constellation. The ancient Egyptians based their calendar on the rising of Sirius, which marked the beginning of the sacred year. As soon as it could be seen, people living near the banks of the Nile would make preparations to relocate to higher ground, realizing that the flood would not be far behind.

FLOOD MYTHS

It would be no exaggeration to say that almost every culture has a flood myth. While the biblical flood epic of Noah and the Ark is perhaps best known to the Western world, ancient civilizations as diverse as Babylonia and Wales, or Russia and Sumatra, have their own versions of a giant flood. Many of these myths are surprisingly similar. The flood is seen as a punishment for mankind's sins, a family is singled out for rescue by a deity, a boat is built, and creatures of all kinds are collected to ride out the storm. The rains that sent Noah on his journey famously lasted 40 days from the time "floodgates of the heavens" and waters from the "springs of the great deep" surged over Earth's surface. Eventually the ark came to rest on the summit of Mount Ararat (which is in present-day Turkey).

After Noah, the best-known hero of the flood sagas is the Babylonian hero Gilgamesh, whose account is enshrined in the *Epic of Gilgamesh*. In ancient Sumer, the man chosen for survival was Ziusudra (meaning "he saw life"). The god Enki warned him of the gods' decision to destroy mankind in a flood. His tale is found in the *Eridu Genesis* that dates back to the seventeenth century B.C.: Ziusudra is instructed to build a large boat. The flood persists for seven days, at the end of which, after making sacrifices to the gods, Ziusudra is rewarded with the gift of eternal life.

The correspondences between the sagas of Noah, Gilgamesh, and Ziusudra are so striking that historians believe that they all emerged from the same oral tradition. There is some tantalizing archaeological evidence that an actual catastrophe may have formed the basis for the proliferation of these myths. Excavations in Iraq have shown evidence of a flood that took place about 2900 to 2750 B.C. But flood myths have arisen independently thousands of miles (kilometers) from the Middle East. In ancient China the gods brought a flood on humans that lasted 22 years as punishment for their wayward behavior. The Mayan and Aztecs had similar flood myths.

Noah and the Ark
The story of Noah and the Ark, chronicled in the Book of Genesis, is only one of about 500 flood myths that have been recorded around the world.

While scholars and historians dispute the cause of the flood described in the Bible, archaeological evidence does suggest that a catastrophic flood, possibly precipitated by an earthquake or a volcanic eruption, did occur thousands of years ago in the Mediterranean basin.

St. Margaret's flood is proverbial and considered to be well for the harvest in England.

The saying reflects the belief that flooding—if it is not catastrophic—can be beneficial because it can help fertilize the land once the waters recede. The most common kinds of flooding come from rivers or streams that are swollen by heavy rainfall or rapid melting of snow. Because St. Margaret's Day is celebrated on June 10, the flood would come at the height of the planting season and ensure a bountiful harvest. Flooding cannot be avoided completely, no matter what precautions are taken, and in any case they are not necessarily advisable.

Some areas are expected to flood periodically—the banks of the Nile, most famously—and are known as flood plains. Flood plains are defined as areas surrounding rivers and creeks where occasional flooding occurs. Far from spurning such areas, people have settled on flood plains since the beginning of agriculture.

DROWNED CITIES

No chamber of commerce is needed to promote Atlantis or put it on the map—that is, if anybody could figure out where it was before it sank, or for that matter, whether it even existed. Atlantis was first mentioned by the great Greek philosopher Plato (c. 428–347 B.C.), who declared that it had disappeared under the Mediterranean 9,000 years before his time. Atlantis sank in the waves "in a single day and night of misfortune," he wrote, because of a natural disaster. Since Plato, historians and archaeologists have argued over its existence, and novelists and filmmakers have made it into a symbol of a lost Eden. Many locations have been proposed for the doomed island—Sardinia, Crete, Santorini, Cyprus, and Malta among them. Some archaeologists have advanced the theory that Atlantis was swept away by the eruption of the volcano on Thera in the seventeenth century B.C., which triggered a tsunami in the Mediterranean. It has been suggested that this catastrophe destroyed the Minoan civilization.

While indisputably the most famous of all drowned cities, Atlantis is by no means alone. In the folklore of Brittany, Ys, once a burgeoning seaport, was submerged as a punishment for the wicked behavior of its inhabitants. Irish folklore has a similar tradition of cities lost underwater. The lake known as Lough Neagh in County Armagh is said to cover one such city. According to a legend first recorded in the twelfth century, the town was destroyed because of the behavior of its residents, who were "sunk in vice, and (were) more specially incorrigibly addicted to the sin of carnal intercourse with beast more than any other people of Ireland." There are tales of fishermen who in calm weather were supposedly able to glimpse the "ecclesiastical towers" of the wicked town sticking up underwater.

The reason is very simple: Flooding makes the soil more fertile and it also provides a readily available supply of water for irrigation long after the flood has ended. When floods recede, they leave behind a layer of nutrient-rich silt on the soil. They can also replenish the groundwater and refresh wetlands. In many parts of Asia, floods enhance rice production. The land where rice is grown often has permeable subsoils composed of clay, which reduce infiltration and water loss. Geologically, flood plains are also congenial locations for erecting buildings, highways, and railroads because they are flat. For many people who make flood plains their home, the benefits outweigh the risk of finding their basements filling up with water from time to time. Flooding is much less desirable in coastal regions, where it is due mostly to storm surges. Floods are not an entirely natural phenomenon; humans have to share some of the blame for more frequent and intense floods in many parts of the world. Removing water-absorbing vegetation near rivers or streams and replacing it with different species, for instance, may inhibit the ability of floodwater from infiltrating into the ground, where it can replenish groundwater. With nowhere else to go, the water only builds up on the surface, making the flood worse. Overgrazing by livestock, deforestation, mining, and urbanization can also contribute to the problem.

Don't sleep in a place so low
that a flood can carry you away.
Don't sleep in a place so high
that the wind can blow you away.

This saying, which comes from the Central Asian nation of Turkmenistan, clearly emphasizes the importance of caution. In places where flash floods occur, this warning is especially warranted. In some parts of the world—in Florida or in parts of southern California, for instance—rainfall can be sudden and torrential. Flash floods have been known to produce 20 feet (6 m) of water in just minutes. Because such a large accumulation of water piles up in such a short time, it can create a powerful force. These floods have been known to sweep cars into their currents and bring down bridges. For people caught off guard, floods can be fatal. In 1878 a freight train, consisting of 10 ox wagons, was washed away by a flash flood resulting from the overflow of Rapid Creek, South Dakota. On June 9, 1972, an intense thunderstorm inundated Black Hills, South Dakota, in 15 feet (4.5 m) of water; as nearby rivers overflowed, they submerged much of Rapid City, drowning 238 people and damaging over 1,300 houses.

Drowned City

There are many legends about submerged cities being under water. Accounts of doomed cities can be found in the United Kingdom, Ireland, France, Spain, and Germany. But underwater archaeologists have discovered that some stories of drowned cities are based on fact. In 1959, for example, skin divers located the ancient Etruscan seaport of Pyrgi off the Tyrrhenian seacoast just north of Rome. While the city was known from historical records, almost no evidence for it had turned up previously.

Frog Folklore

The 5,000 species of frog are found virtually all over the world and in almost every type of terrain; some species have been reported at 13,500 feet (4.1 km) high on the Tibetan plateau.

Frogs also serve as indicators of the health of the environment and recent mysterious die-offs of certain species have raised alarms among scientists. Frogs have long been the object of folklore. The ancient Chinese believed that frog spawn fell from heaven. In ancient Egypt the frog was associated with Heqet, the goddess of fertility.

An April flood carries away the frog and his brood.

Frogs like water—that's hardly a secret. But in fact, while frogs spend much of their lives in marshes, in swamps, or along a river, they don't always stay in water for long periods of time except when it comes to producing a brood. It doesn't necessarily have to be a lake or a marsh; a puddle after a heavy rain will do just fine. So breeding often begins in earnest after heavy rains. Frogs typically lay their eggs in water, although some species lay their eggs in soil, beneath logs, or on leaves. The breeding season begins once frogs emerge from hibernation in February and early March, although they can breed as early as December or as late as April.

Frogs generally return for breeding to the sites where they originally developed into mature frogs the previous year. Males arrive first and begin to croak—a low purring croak, as it has been described—to lure a mate. Females will lay 1,000 to 4,000 eggs at a time in shallow, still water. The eggs are called frog spawn, and they're surrounded with a clear jellylike substance that swells up in the water to protect the embryos. The spawn float to the surface in large round clumps so that they can take advantage of the warmth of the sun. It takes 30 to 40 days for a tadpole to form and three more years before frogs are sexually mature. While rainfall is welcome prior to the breeding season, too much rain after the eggs are already laid can be devastating for the amphibians. The spawn, already vulnerable to predators, can be swept away by flooding, and so can the parents.

Frogs and floods also figure in the weather lore of native Australians. Like many peoples around the world, the Aborigines have a flood myth. In their myth the flood is perpetrated on mankind by a frog: "The frog opened his sleepy eyes, his big body quivered, his face relaxed, and, at last, he burst into a laugh that sounded like rolling thunder. The water poured from his mouth in a flood. It filled the deepest rivers and covered the land. Only the highest mountain peaks were visible, like islands in the sea. Many men and animals were drowned."

A foggy day indicates floods.

While foggy days will not necessarily indicate floods, fog and floods do have an interesting connection. In the summer of 1783 a "constant fog" hung over Europe and parts of North America. "This fog was of a permanent nature; it was dry, and the rays of the sun seemed to have little effect towards dissipating it," wrote

Benjamin Franklin. Because solar radiation was so weak, the ground remained frozen long after it should have thawed, and the winter was much colder than normal. The unusually protracted fog was later traced to an Icelandic volcano named Grimsvotn. There is a glacier lake in the middle of the volcano that periodically produces floods, known as the *jokulhlaups* ("glacier's run"). Whether or not the volcano erupts, the glacier bursts about once every 5 to 10 years, flooding the area.

WHY FLOODS OCCUR

Floods occur for many reasons. Along rivers and streams, flooding is usually precipitated by heavy rainfall or by an unexpectedly large runoff from melting snow in spring. In coastal regions, storms are more likely to cause floods. Hurricanes and typhoons, for instance, are responsible for catastrophic flooding that can reach far inland. Where inland rivers wind through hilly or mountainous areas or where rivers drain into coastal waters, flooding can occur more quickly because the rivers are steeper. However, when flooding happens in this way it is usually brief—often lasting for only one to two days. Floods are also caused by or exasperated due to human action or inaction: Overgrazing, overdevelopment, mining, and deforestation can all contribute to making a bad problem worse.

When flash flooding occurs, it's usually the result of a short intense rainfall, commonly from thunderstorms. Flash flooding is a particular danger in urban areas, especially if drainage systems are unable to cope with a large flow of water in a short period.

Flash Flood in the Mountains

These downpours usually result from thunderstorms. The rain descends so quickly that the mountainous soil is unable to absorb it, with the result that it rushes down the slopes into the valley, overwhelming anything in its path. The destruction may be even greater if the surge of water is diverted or blocked by an obstruction such as ice or a dam. Flash floods often occur in regions that experience precipitation infrequently, so people may be even more unprepared for them.

5

STORMS, WINDS, AND TORNADOES

Captain George Nares, a nineteenth-century Scottish naval officer and polar explorer, was always on the lookout for hurricanes. "June—too soon," he wrote. "July—stand by; August—look out you must; September—remember; October—all over." Whatever the type of storm, hurricane, typhoon, or tornado, two things can safely be said: It will be unpredictable, and it will be destructive. Efforts to anticipate when storms will strike have informed weather folklore no less than science.

STORMS

"And another storm brewing," William Shakespeare wrote in *The Tempest*. "I hear it sing in the wind." Storms may not sing, but they make themselves heard. Storms are still brewing, and because of the effects of suspected climate change, they may be growing stronger in places. They are unpredictable; weather folklore is an attempt to find a natural order in phenomena that seem to defy it.

If St. Elmo's Fire be single, prognosticates a severe storm, which will be much more severe if the ball does not adhere to the mast, but rolls or dances about. But if there are two of them, and that, too, when the storm has increased, it is reckoned a good sign. But if there are three of them, the storm will become more fearful.

This saying from the Roman author and naturalist Pliny the Elder (A.D. 23–79), by way of the English philosopher and statesman Francis Bacon (1561–1626), refers to a phenomenon resulting from a coronal discharge caused by the ionization of the air during thunderstorms within a strong electric field. It appears as a blue glow that is often seen at sea on ship masts (St. Elmo is the patron saint of sailors); or on spires and chimneys; or on the wings, tail, or nose of aircraft. Generally, St. Elmo's fire is observed when severe thunderstorms are dissipating, but it may also indicate conditions in which more violent storms can occur.

Howling Wolf

Wolf tales are common to many cultures and extend back centuries. Sometimes portrayed as beneficent and helpful to humans, wolves have also been seen as evil incarnate. In a Scottish legend, killer wolves ate children. China had its own mythically murderous wolf called Lon Po Po. And, of course, no wolf tale is better known than the story of Little Red Riding Hood.

Wolves always howl more before a storm.

Wolves may howl more before a storm; although the evidence is scant. They can also whimper, whine, growl, bark, or yip. Indeed, while a howl is the vocalization most frequently associated with wolves, why they howl is a more complicated question. They may howl because they are happy, to communicate with others in their pack, or to assert territorial rights and keep other packs away. They also seem to do a group howl, possibly as a way of promoting solidarity. One researcher observed a pack of arctic wolves howling in different tones, more like individual members of a symphonic orchestra. He speculated that this was an evolutionary mechanism to give potential predators the impression that the pack was larger and more formidable than it actually was.

Dogs, which owe their origin to wolves and the humans who bred them for favorable traits, do not seem to bark more before storms at all. Some dog lovers have observed that they are inclined to bark more frequently at night after a hot day. It is possible that having napped during the height of the day's heat, they are more active and restless at night. Because the weather allows them to be outdoors in warmer weather, they also may not need to take their masters' sensibilities into account as much as if they were housebound.

If the rooster crows on going to bed,
you may rise with a watery head.

The decrease in atmospheric pressure, which may signal an approaching storm, is believed to especially affect birds and may account for a sudden spurt in restless behavior. In that case a rooster might react in response by crowing in the middle of the night rather than waiting until dawn.

OWLS AS WEATHER FORECASTERS

Owls have had a sinister reputation through no fault of their own. In the nineteenth century, the British poets Robert Blair (1699–1746) and William Wordsworth (1770–1850) both referred to the humble barn owl as "the bird of doom," apparently because these nocturnal creatures were associated with darkness. Besides announcing the impending death of a sick person, owls have also been known as weather forecasters. If, for instance, an owl hoots on the east side of a mountain, bad weather is coming. If bad weather had already arrived, and an owl screeched, that meant that you could expect a change in the weather. And if you wanted to keep away lightning (and ward off evil as well), you were advised to nail a barn owl to the door of a barn, a custom that persisted into the nineteenth century. But more recently, barn owls have been more the victim of weather change than a predictor of it. In 2006 alarmed conservationists reported that the number of barn owls in England had fallen from 4,000 breeding pairs in the previous year to only 1,000. While they held out hope for recovery of the population, they ascribed the dramatic decline to climate change.

Cumulonimbus Clouds

These clouds are associated with violent weather—hail, thunderstorms with torrential rain, and tornadoes.

I know ladies by the score,
whose hair foretells the storm;
long before it begins to pour,
their curls take a drooping form.

Human hair, especially blond hair, can serve as a barometer because it has a tendency to expand in length as the humidity rises. So more humidity in the air, which may portend a storm, may indeed cause naturally curly hair to droop or cause straight hair to curl up a bit. Traditional hygrometers, used to measure humidity, relied on human or animal hair for the same purpose.

A coming storm your shooting corns presage,
and aches will throb, your hollow tooth will rage.
If your corns all ache and itch,
the weather fair will make a switch.

Many people, especially those with migraines and joint problems such as rheumatoid arthritis, are convinced that an intensification of pain is an indication of an approaching storm. But is this true, and if so, can it be scientifically

STORM FORMATION

Simply defined, a storm is any disturbance in the atmosphere that has an affect on weather conditions on the ground. Storms typically arise from the center of a low-pressure region that is surrounded by a high-pressure system. Fronts develop at the boundaries of the opposite air masses because they have different temperatures: Instead of the air masses mixing, they move up and over each other. In addition to high winds, storms manifest themselves in moisture-laden clouds, heavy precipitation (ranging from rain and snow to sleet and hailstones), and thunder and lightning. They can also produce a variety of other violent weather, from tornadoes to whirlwinds. Low-pressure regions develop when packets of warm air rise into the atmosphere; the water vapor in these packets begins to condense and form clouds such as cumulonimbus.

What type of storm forms will depend on many factors. Thunderstorms, for example, develop in warmer weather—in spring and summer in the Northern Hemisphere—while blizzards are typically a winter phenomenon, though they can occur in late fall or early spring. Climate also has a great deal to do with the type of storm that is seen; most of South

proven? Researchers have grappled with this problem for years without coming to a firm conclusion. However, they have collected intriguing evidence demonstrating a connection between pain and air (barometric) pressure. The connection may be due to the fact that a drop in atmospheric pressure causes blood vessels to dilate slightly, aggravating already irritated nerves, as is the case with corns and arthritic joints. In 2003 Japanese researchers published a study that showed a direct connection between low pressure and low temperatures and joint pain in rats. But whether that same effect happens in humans is more difficult to establish.

When asked to keep logs, many people with migraines turned out not to be as affected by storms as they had imagined. Complicating matters, some of them were more sensitive to high temperatures and high humidity, while others experienced more intense migraines under opposite conditions. Dr. John Parenti, director of the Department of Orthopaedics at Geisinger Medical Center in Danville, Pennsylvania, is willing to concede the correlation between pain and low pressure, but added, "I wouldn't base any kind of treatment on it." Americans eager to avoid low-pressure systems because they suffer from more aching joints as a storm threatens are advised to move to Hawaii or northern California, both of which experience fewer severe storms than any other location in the United States.

America and Africa experience little or no snow except at higher elevations (in the Andes, for instance). Dust storms are phenomena that arise in drier climates. Cyclones, such as hurricanes and typhoons, form in warm waters (of the Atlantic and Caribbean on the one hand, and in the Pacific on the other) in summer and fall. They are characterized by a closed circulation around a center of low pressure, and they're kept fueled by heat released by the rising and condensing of moist air from the water. Hurricanes and typhoons are formed by depressions, or "lows," which bring rain, strong winds, and changeable conditions. Depressions often travel from west to east, starting in the Atlantic Ocean. The name underscores their origin in the tropics and their cyclonic nature.

Each type of storm is classified in terms of its precipitation, winds, and temperature. A blizzard, for example, must have gale-force winds—39 to 46 m.p.h. (62 to 74 kph) or greater—heavy snow, and very cold conditions. By contrast a snowstorm, while capable of producing similar accumulations, is not accompanied by high winds. There are more unusual storm types such as windstorms, which do not have any rain, and dust devils, defined as "a small, localized updraft of rising air." Tornadoes are regarded as the most destructive storms of all.

Contrails
Condensation or vapor trails form at very high altitudes (above 5 miles or 8 km), where the temperature is extremely cold.

A storm moderates, to storm again.

While there can be no definitive reading to this saying, it is possible that it refers to a hurricane or other cyclonic storm. Hurricanes have "eyes"—regions of calm in the middle of the swirling winds and intense rains. When the eye passes over a region, the storm will appear to abate; but once it moves on, the other side of the storm passes over, delivering another punishing blow.

A curdly sky will not leave the earth long dry.

This saying refers to high clouds that are composed of ice. In fair weather, these clouds are said to resemble buttermilk spilled on a glass surface. In England these clouds are called "clabbered sky" (from clabbered milk, thin milk like yogurt). Buttermilk skies often presage a large-scale change within a day or two. In the morning this is seen as a benign indicator, but later in the day it may

CONTRAILS AS WEATHER INDICATORS

Contrails (condensation trails) are man-made clouds, formed by the exhaust of jet planes. They are easily identifiable as long, white plumes. Contrails also serve as indicators of stormy weather. On clear dry days, jets will leave no trail, or a trail that quickly vanishes. That is read as a sign that dry weather will continue. However, if the wispy, white trail remains for some time in the sky on a clear day, then rain, snow, or some other form of precipitation may be on the way.

Scientists believe that contrails form larger cloud banks that substantially alter the atmosphere's heat balance. Large contrails only form at an atmosphere where conditions are sufficiently moist and cool—typically somewhere in the range of -40° to -85°F (-40° to -65°C). Over the southern United States, for example, contrails seldom develop because conditions are too hot or too dry.

It is believed that contrails may play a role in altering weather conditions. An unplanned experiment measuring the difference occurred after the terrorist attacks on New York and Washington, D.C., on September 11, 2001. For the next three days, virtually all commercial air traffic over the United States was grounded. Climatologists discovered that variations in temperature were greater in their absence; they were 2°F (1.1°C) higher and lower than normal. That's because contrails act in the same way that cirrus clouds do by blocking out solar energy from above and trapping heat below. The cloud covering contributes to narrowing the daily range in highs and lows.

predict approaching storms. But milk can curdle and so can the clouds—at least metaphorically. So a curdling sky is an indicator that warm winds are rising up to high levels in the atmosphere under calm conditions.

When forests murmur and the mountain roars, then close your windows and shut your doors.

Winds tend to begin at higher elevations—mountaintops, for example—and then descend to Earth. When a cold front associated with a storm approaches, the air current will stir the forest. Meanwhile, on the top of the mountains, the winds will have begun to roar.

Evangelista Torricelli
The barometer was invented in 1643 by the Italian physicist and mathematician Evangelista Torricelli. The "torr," a unit of pressure, was named in his honor for his work with mercury.

BAROMETERS

There are few more important weather markers than atmospheric pressure, because it presages a change in weather conditions. The instrument most commonly used to measure atmospheric pressure is a barometer. The word comes from the Greek words for weight and measure. Invented by the Italian physicist Evangelista Torricelli in 1643, the barometer is used to tell whether a storm is approaching or good weather is in store. The difference in air pressure can cause winds to blow, with air moving from an area of high pressure to an area of low pressure. As a result, barometers are also a useful tool in predicting the direction and strength of winds.

Although there are different types of barometers, they all are based on the same basic principle: measuring the weight of a vertical column of air through the atmosphere. Essentially, the barometer is a glass tube from which the air has been removed, inserted into a dish of mercury. The mercury responds to the pressure of the air pushing down on it by moving up into the vacuum of the glass tube. When pressure decreases, it means lower pressure, which can in turn spell bad weather ahead. There are three basic types of barometer: mercury, aneroid, and digital. The most accurate barometers (and not coincidentally, the most expensive) are digital barometers, which use electronic circuitry to detect changes in pressure.

GUST FRONTS

Gust fronts announce the approach of a thunderstorm. They form the leading edge of rain-cooled air that descends from a thunderstorm. Their passage is accompanied by a shift in wind direction and a sudden drop in temperature—as much as 10°F (6°C) within a few minutes.

Gust front winds can flow outward at speeds of up to 100 m.p.h. (161 kph). Known as outflow boundaries (between the cooler and warmer air masses), they can persist for up to 24 hours after the thunderstorm has dissipated. They can travel hundreds of miles (or kilometers) from their source, creating new thunderstorms when they run up against other air masses.

A weathercock that swings to the west proclaims the weather to be the best. A weathercock that swings to the east proclaims no good to man or beast.

This old saying refers to any wind from the southwest that swings around the compass to the northwest. Such winds generally bring dry weather. Winds from the east are more likely to bring rain. Shifts in wind tend to be more abrupt and more frequent in New England (six northeastern states in the United States) than in Europe. Where such shifts are more dramatic and more frequent, a wind-measuring device, such as a traditional weathercock (or weather vane), takes on added importance when it comes to predicting future weather conditions. For early American farmers, the weathercock was an essential tool. The first weather vanes were made from light pine or cedar and could swing with even the slightest change of the wind. The more familiar ornamental weathercocks—sporting roosters, whales, or horses—came later.

As the days grow longer, the storms grow stronger.

Rhyme trumps truth in this saying—for while it is true that as days grow longer in spring, thunderstorm activity becomes more frequent due to rising temperatures and greater humidity, it is not uncommon for fierce wintry blizzards to strike on short, dark winter days. And as the days grow shorter in late summer and in early autumn—especially in the United States, Mexico, and the Caribbean—hurricanes can cause immense devastation.

When the sun draws water, storms will follow.

The sun does not draw water. This saying describes an optical illusion in which the sun's rays alternate with bands of shadow to produce a fanlike effect. Those shadowy patches are dense clouds, some of which are thin enough to allow sunlight to reach Earth. However, the saying is not without some merit. If the sun is obscured in the west, it means that moisture-laden clouds have gathered there, and it is quite possible that rain will follow if the temperature is favorable for the condensation of that moisture.

Eye of the Storm

The most destructive storms arise from a cyclonic system—an area of low atmospheric pressure typified by spiral winds and often accompanied by intense precipitation. In the Northern Hemisphere, the winds rotate counterclockwise, and in the Southern Hemisphere they turn clockwise. Cyclones can produce a variety of storms including hurricanes, typhoons, and tornadoes.

THE SUPERSTORM

Some storm systems not only pack a wallop, but can produce a bizarre variety of catastrophic weather. This is what happened in the so-called "superstorm" of March 1993 in the United States, when the eastern third of the country was pummeled with storms ranging from blizzards to tornadoes, all generated by a single system.

Hurricane winds buffeted the region from Alabama to New England, which also experienced record cold and over 40 inches (102 cm) of snow. Nearly every major interstate highway was shut down (and secondary roads made impassable) and for the first time, a storm forced every major airport on the East Coast to close. Millions of people lost electricity and the storm claimed 270 lives, while causing damage totaling $3 billion.

Lightning Strike

Lightning can cause a great deal of damage to buildings as well as kill and injure people caught in storms. In the United States, for instance, 3,239 deaths and 9,818 injuries from lightning strikes were recorded between 1959 and 1994. Only flash floods and river floods cause more weather-related deaths. Curiously, men are four times more likely to be hit by lightning than women.

THUNDER AND LIGHTNING

People hit by lightning were thought by many ancient Africans to have angered the gods. The Greek god Zeus was never without his thunderbolt. Bolts could literally come out of the blue when the clouds yielding them could not be seen, and the destruction they could cause has inspired many stories, legends, and folklore. Thunder and lightning may no longer be the preserve of a Zeus, but they exert a hold over our imaginations that explanations of positive and negative charges and corona discharges can never match.

Thunder without a cloudburst is indeed cursed.

Although thunder is generally associated with rain (and lightning), some thunderstorms come without cloudbursts—that is, they do not produce any rain. This is known as a dry thunderstorm. These occur more frequently in regions where the air near the surface is often desert-dry and where the base of the cumulonimbus clouds, which typically bring rain, are unusually high. It is not true to say that the clouds release no rain; instead it evaporates in the dry air before it can reach Earth. Even hail can melt and evaporate before reaching the surface. However, a dry thunderstorm will still produce lightning and cause gusty, shifting winds. A dry thunderstorm evokes dread because while the lightning can trigger fires on the ground, there is no rain to put them out or mitigate their impact.

Thunder in the morning, all day storming.
Thunder at night is the travelers' delight.

This saying reflects the observation that a thunderstorm in the evening will probably be gone by the time the next day dawns. Thunder in the morning might mean that conditions during the night had changed to favor precipitation for an extended period, with cooler temperatures on the ground and an accumulation of moisture-bearing clouds aloft. In most cases, however, thunderstorms do not take place until later in the day on warm spring and summer afternoons. That is because a sufficient amount of moisture must rise into the atmosphere, something that usually can occur only if the ground heats up from solar radiation over several hours. During the night, temperatures on the ground are usually too cool for that to happen, unless the ground retains enough heat that is built up through the day.

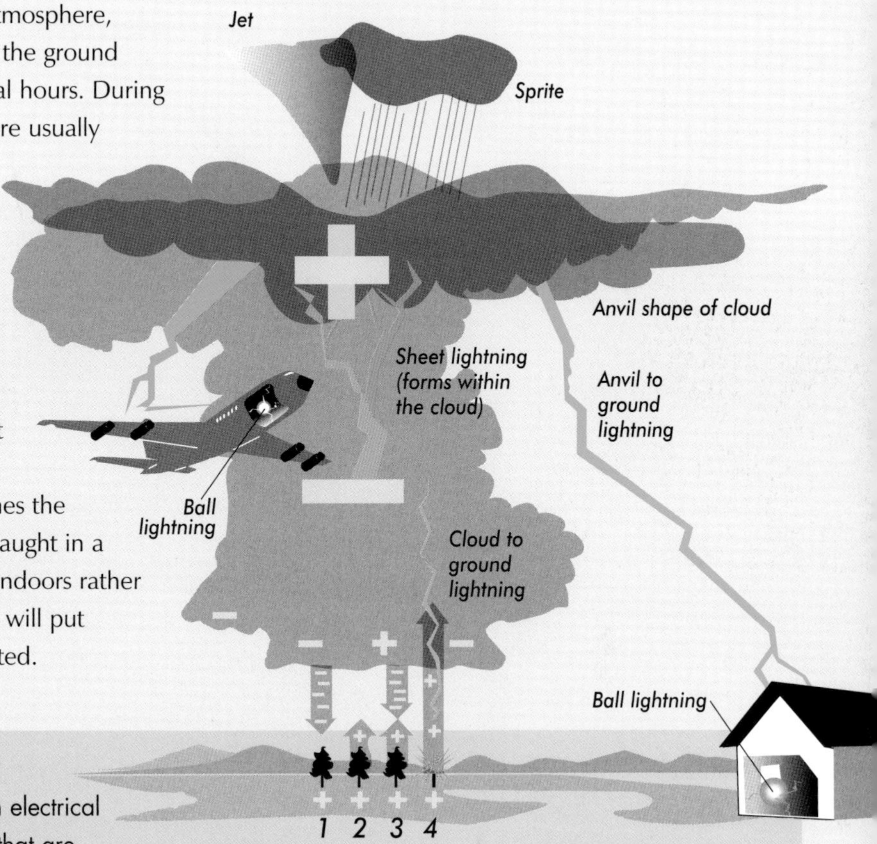

Lightning strikes more trees than blades of grass.

This German saying simply indicates that lightning is attracted to higher objects, and so it will strike a tree before it reaches the ground. That accounts for why people caught in a thunderstorm are urged to seek shelter indoors rather than take cover under a tree, since they will put themselves in danger of being electrocuted.

LIGHTNING FORMATION

Lightning can be simply defined as an electrical discharge from cumulonimbus clouds that are responsible for thunderstorms. They can be found at altitudes as high as 60,000 feet (18 km) above the Earth. Conditions in these clouds are turbulent, meaning that there is a lot of movement of air. That movement causes large differences between electrical distribution, with the top of the cloud becoming positively charged and the bottom negatively charged. That is because ice crystals that are pushed to the top of the cloud due to powerful updrafts have a positive charge, while the hailstones, which are pushed to the base by the downdrafts, have a negative charge. A positive charge also develops on the ground. (Under calm conditions there is no difference in charge; a calm sky would be neutrally charged.)

Lightning usually originates from a cloud and strikes the ground, but it can also travel between two clouds. During a thunderstorm, ice crystals with a positive charge are pushed to the top of the cloud while hailstones with a negative charge are pushed in the opposite direction. The negatively charged hailstones are attracted to the positively charged ground (1) or to tall pointed objects such as steeples or antennas (2). The two charges meet (3) and establish a balance in an upward transfer of positive charges (4).

HOW FAR AWAY IS THE STORM?
It is fairly simple to estimate how far away a thunderstorm is. Lightning will always be seen before you hear thunder because the speed of light is much greater than the speed of sound.

Once you see a flash of lightning, count off the seconds until you hear the thunder. Divide the number of seconds by 5. The result gives the approximate distance from your position to the thunderstorm. The rule is that 5 seconds is the equalivalent of 1 mile (1.6 km).

A more precise calculation requires a consideration of other factors such as humidity, pollution, and heat, all of which can affect the ability of sound to travel.

Thunderstorms in June mean the grain will not be lean.

This saying is both true and false. Cool, rainy weather in early summer may be favorable for small grain and hay crops, but can retard the growth of corn. Thunderstorms may be generated by warm, humid conditions, but are often followed by cold fronts. Farmers prefer precipitation early in the growing season, when it can do most good, rather than later, when it can disrupt harvesting.

Beware the bolts from north or west; in south or east the bolts be best.

Stormy weather often comes from the north or west, so lightning seen in either direction might mean that a storm is approaching. However, as storms dissipate, they will tend to move in the opposite direction; so if lightning is seen in the south or east, the observer can be assured that the storm is moving away and that he or she has little to fear. Generally speaking, winds blowing from the southeast, northeast, and north are likely to bring steady rain or snow. More pleasant weather may be en route when the winds are blowing from the west to the northwest.

When cumulus clouds become heaped in leeward during a strong wind at sunset, thunder may be expected during the night.

Cumulus clouds are typically fair-weather clouds. However, as winds shift, cumulus clouds can develop vertically, reaching high into the atmosphere, where they can form cumulonimbus clouds that are full of moisture and involved in violent thunderstorms.

When March blows its horn, your barn will be filled with hay and corn.

The phrase "March blows" refers to thunderstorms. The saying means that such a March is unusually warm. (In some rural parts of the United States, annoyed mothers would exclaim, "Thunder in the wintertime!" when their children misbehaved.) Thunderstorms can occur during warm weather only when there is

THUNDER GODS

Of all the thunder gods of mythology, perhaps the best known is Thor, the Norse god, son of Odin, whose magical hammer was guaranteed to come back whenever it was thrown—much like a boomerang. This hammer, for which he needed a pair of iron gloves, was capable of causing thunderclaps as well as killing giants. A hammer or some other instrument of power and an unusual means of conveyance both seem de rigueur for thunder gods. In Greek mythology, Zeus, besides his responsibilities of presiding over the other quarrelsome gods, also held the post of thunder god, hurling thunderbolts from his home on top of Mount Olympus. The Hurrians, an ancient people that inhabited the Near East, honored Teshub, the god of sky and storm. He was often shown with a triple thunderbolt and an ax or mace. In ancient China, the thunder god's role was assumed by Lei Gong, who began life as a mortal. Many North American tribes venerated thunder gods. The Sioux deity, Haokah, had a contrary personality: He would weep with the rain in happiness, and in grief, smile with the sun. He used his great drum to beat out thunderclaps; his drumstick was the wind. He, too, would fling thunderbolts down to Earth.

a large difference between the temperature on the ground and the temperature in the atmosphere. Warm air rises from the ground, and if it cools sufficiently in its ascent, it may condense and form precipitation. That is why most thunderstorms tend to occur in summer. However, only someone of a very optimistic bent would rush out to plant in March. It is human nature to think that trends will persist indefinitely. If the stock market has been going up for months, it will continue to ascend meteoric heights; if it has been raining for 40 days and nights, then you could be forgiven for thinking that it will rain the next day, too. But for anyone with a long memory, spring is too fickle a season to make any bets on.

Thor's Hammer

Thor, the Norse god of thunder, was usually depicted with a hammer in his hand (which he used to produce thunder). He was a big, shaggy-haired figure with eyes of lightning who was revered as a protector of humans against evil.

A pinching crawdad will hold on until it hears thunder.

A crawdad (also known as a crayfish, spoondog, or yabbie, if you are Australian) is a freshwater crustacean found in many rivers and streams in North America, Europe, and parts of Asia. While it is debatable as to whether they will cling to an object (or a quarry) until a thunderclap, as this saying from Appalachia declares, it does have powerful claws, which can make short work of snails and small fish (which constitute their diet) and are capable of shearing a person's finger clean off.

Lots of low rolling thunder in the late fall means a bad winter.

"On St. Lucius' Day (October 26, 1253), there fell a great snow," wrote a British historian, "and with all the winter's thunder, for a token of some evil to follow." Thunder during snowstorms is a result of a buildup of static electricity in the air, along with atmospheric instability. The phenomenon may be caused by static electrical discharges that occur with thunderstorms. Sometimes lightning (or something that resembles it) can occur during snowstorms as well. In 1964 meteorologists in Tucson, Arizona, witnessed an intense series of light flashes at or near the ground in the midst of a snowstorm that took place without any thunder. They speculated that the heavy snowflakes were dragging down the discharges from the clouds.

UNUSUAL LIGHTNING

While there are many forms of lightning (forked and sheet being the most common), few are more bizarre than ball lightning, a short-lived, glowing, globe-shaped phenomenon that can hover over the ground. It is occasionally associated with thunderstorms; but unlike ordinary lightning, which flashes and vanishes in an instant, ball lightning can persist for several seconds. There have been reports of ball lightning in good weather, too. Ball lightning has been documented since the seventeenth century, but how it is formed remains unclear. Scientists are still in dispute as to the cause, and there is some speculation that several different phenomena have been mistakenly lumped together. Ball lightning has a tendency to float in the air and may appear red, yellow, or white, or even transparent.

St. Elmo's fire is not a form of lightning at all, though it can be seen in thunderstorm activity. It is defined as a "glow discharge," a continuous electric spark that glows in a similar way to the glow from a fluorescent tube. St. Elmo's fire can appear in the middle of thunderstorms under two conditions: the ground has to be electrically charged, and the air between the cloud and the ground has to have a high voltage. That voltage rips apart the air molecules, separating the protons from the electrons. The charges stream off the ionized molecules and produce glowing discharges—technically known as a "corona discharge."

St. Elmo's fire was often observed by sailors at sea who regarded it as an omen of heavenly intervention. Its name is supposed to come from the Italian derivation of Sant 'Ermo or St. Erasmus (c. A.D. 300), the patron saint of the early sailors who plied the Mediterranean Sea. References to the phenomenon have been recorded since antiquity.

Ball lightning

A rare phenomenon, ball lightning can assume the form of a ball and seem to hover for several seconds in the air before disappearing.

Although it has been associated with thunderstorm activity, it is not formed like ordinary lightning that jumps from one point to another and vanishes in a moment. This artwork from 1900 captures when the German physicist Georg Richmann was killed by ball lightning in 1753.

When the flashes of lightning appear very pale, it argues the air to be full of waterish meteors; and if red and fiery, inclining to winds and tempests.

Lightning emits white light as it streaks through the air, but it can appear in different colors, depending on local atmospheric conditions. It is similar to sunlight in that respect. Sunlight is white light, but can assume different colors, especially at sunrise and sunset, because of the distorting effects of moisture, haze, dust, and pollution in the lower levels of the atmosphere. Shorter wavelengths will often be scattered out—greens, blues, and purples—while the longer wavelengths (the reds and oranges) will dominate. For these effects to be visible to the observer, lightning has to take place at some distance. The color of lightning will also change just before it strikes an object on the ground, so for the final 10 feet or so (3 m) above the ground, it may appear as a bright, fiery orange or red. When there is a great deal of moisture in the air, as is the case during a thunderstorm or over oceans, lightning will typically appear yellow. So the saying seems weak in terms of its ability to predict the type of weather by the coloration of lightning.

If there be sheet lightning with a clear sky on spring, summer, or autumn evenings, expect heavy rain.

The flashes of light across the horizon do have a sheetlike appearance. Sheet lightning, as it's called, is not lightning itself as much as it is the reflection of lightning on parts of a thundercloud seen from a distance. However, because of the distance, individual lightning bolts cannot be viewed. These flashes may be the only evidence that a thunderstorm is actually occurring, since the rainfall is too far away to be observed, the thunder too far away to be heard, and the winds too far away to be felt. Sheet lightning has also been called heat lightning because on hot nights, small thunderstorms can be localized; that means that some people will be aware of a storm only by these flashbulb bursts of light. This optical effect gave rise to the false belief that heat itself could generate lightning.

Forked Lightning

Forked lightning does not set the stage for any type of weather conditions the following day. To predict future weather conditions after a thunderstorm, more attention should be paid to the movement of clouds that produced the lightning.

Forked lightning at night, the next day clear and bright.

Forked lightning gets its name because it appears as jagged or crooked lines of light and usually follows a zigzagging pattern. It is also known as branch lightning because it can have several branches. Forked lightning usually travels from one cloud to another cloud, but about 20 percent of the time it will strike the ground. On rare occasions it will discharge from a cloud into space. This lightning can strike up to 10 miles (16 km) away from a thunderstorm, so it can look as if it is coming out of a clear blue sky—literally a "bolt from the blue."

This lightning begins its journey out of the cloud as a stream of weakly charged particles—so weak that it emits very little light. As these particles travel through the atmosphere, they search for the path of least resistance toward oppositely charged particles. That process accounts for the branching. As these branches collide with streams coming in the other direction, they trigger return strokes. These reactions come so quickly that it appears as if all the branches are lighting up.

Lightning never strikes the same place twice.

This is one of the most famous weather sayings—and it's wrong. Lightning not only can strike the same place twice, but it seems to prefer high locations. New York City's Empire State Building, for example, is struck about 25 times every year.

A thunderstorm comes up against the wind.

Thunderstorms usually occur when a cold air mass moves over a region where the air is warmer. Cold air moves more quickly than warm air, so it can be said to overtake the warmer air, lifting it up into the atmosphere—where the water vapor it contains will undergo condensation, form clouds, and return to the ground in the form of precipitation. Thunderstorms do not bring cooler weather. But temperatures after the thunderstorm has passed will depend on the winds that have brought the storm. If the winds originate in the east, temperatures will be moderate; but if another thunderstorm should move in from the west, the two air masses will produce higher winds and lower temperatures. The difference between thunderstorms that come from one direction versus another will depend on many factors, geography and prevailing climactic conditions among them. In the United

LIGHTNING SAFETY TIPS

- Do not shower or bathe during a thunderstorm.
- Take cover indoors. If you are in a car with a hard top, stay inside and roll up the windows.
- Remain a few feet (a meter or so) away from windows, sinks, toilets, showers, electrical boxes, and appliances. Avoid landline phones. Cell phones and cordless phones can be used.
- Do not use metallic objects.
- Avoid trees, sheds, lean-tos, carports, or partially open shelters such as pavilions.
- If you cannot find shelter in time and your skin tingles or your hair stands on end, lightning may be about to strike. Find a low spot away from trees, and metallic, tall, or long objects. You should maintain a distance of at least 15 feet (4.5 m) from other people. Crouch down and bend forward with your hands on your knees. Never lie down. If you are in the water when a thunderstorm is moving in, get to land at once and try to find shelter.
- People struck by lightning will receive a severe electrical shock and might suffer burns. Even someone whose heart has stopped as a result of a strike can often be revived by prompt CPR (cardiopulmonary resuscitation). Do not let victims walk around—give them first aid.
- You should wait for 30 minutes after you see the last flash of lightning before resuming activities outdoors.

States, for instance, thunderstorms coming from the west are often drier due to the more arid regions of the country that the winds pass over. As a result, dry thunderstorms may occur in which rain (or sleet or snow) evaporates before it can reach the ground, although lightning will still occur, along with thunderclaps. A storm from the east, where more temperate and wetter conditions prevail, is less likely to have the same effect.

Lightning in spring indicates a good fruit year.

If there is lightning in spring, then thunder—and thunderstorms—cannot be far behind. Thunderstorms typically occur in humid weather with rising temperatures, conditions that can be favorable for fruit. However, bloom times for fruit can vary from year to year and from location to location. In some places, for example, pear trees always bloom earlier than apple trees do. A sudden shift of weather patterns can also have considerable impact. Unseasonably warm weather, which is implied by the saying, can cause trees to bloom very quickly.

Be Aware

Lightning does not strike just during the height of the thunderstorm. The greatest danger often comes with the first or last flash, when people least expect it. Do not wait for lightning to strike nearby before taking cover.

WHAT IS WIND?
Wind is simply air in horizontal motion, caused by the uneven heating of Earth's surface. Wind acts as a mechanism to compensate for inconsistent temperatures and equalize pressure differences due to variations in temperature. The greater the differences in pressure, the stronger the wind will be.

WINDS

Wind is prophetic in that it brings news of approaching weather. Whether they are light breezes or gales that pack enough force to level towns, winds have inspired poems, songs, and folklore since ancient times. Predicting the velocity and direction of the wind is one thing, but controlling it another; if ordinary human beings could not do it, perhaps magicians could. In an old European legend, magicians managed to capture winds in bags and tied them with ropes. The magicians would then manipulate the winds by the tightness and number of ropes. But those of us without special powers must settle for capturing winds with words.

A high wind drives away the frost.

This saying is not precise, but it does reflect a reality. Wind cannot dispel frost; however, what wind can do is curb the formation of frost. To develop, frost needs cooling temperatures, whereas wind can bring warm air, slowing the cooling process. Moreover, wind can stir up the atmosphere, making it more difficult for radiation to escape into space.

Do business with men when the wind is in the northwest.

Northwest winds generally bring pleasant weather. This saying from Yorkshire, England, suggests that favorable winds and pleasant conditions will improve peoples' dispositions and make them more amenable when it comes time to transact business.

When the wind backs, and the weatherglass falls, then be on your guard against rain and squalls.

A backing wind, which shifts in a counterclockwise direction, portends the approach of a low-pressure system and stormy weather. The weatherglass is a traditional barometer and made its first appearance in the sixteenth century, its invention attributed to Dutch nobleman Gheijsbrecht de Donckere. Early

JET STREAMS

Jet streams are rapidly flowing currents of wind that are found 6 to 9 miles (10 to 14.5 km) above Earth's surface. These meandering currents are typically thousands of miles (or kilometers) long, a few hundred miles wide, and only a few miles thick. Jet streams are responsible for steering storms and influencing the location of areas of high and low pressure. They are created as a result of temperature differences between adjacent air masses—for example, between polar air masses and air masses formed over middle latitudes. Put another way, the collision of air masses of high pressure (warmer air) and low pressure (cooler air) produces winds at high altitudes. There is a total of four jet streams—two main jet streams at each pole and two minor jet streams that flow closer to the equator. The jet streams owe their discovery to the Japanese meteorologist Wasaburo Ooishi in the 1920s, based on observations he made tracking weather balloons.

American colonialists called it the Cape Cod weatherglass (after Cape Cod, Massachusetts) or the thunder bottle. The weatherglass enjoyed considerable popularity with fishermen and farmers through the nineteenth century. A weatherglass consists of a pear-shaped glass bottle filled with water, which may be flattened on one side so it can hang on the wall. It has a long spout (called a tulle) that is sealed except for its end, which is open. The bottle is filled with water and hung in an upright position away from sunlight. Changes in atmospheric pressure will cause the water in the tulle to rise or fall in response. A water weatherglass—in contrast to mercury, aneroid, and electronic barometers—cannot measure absolute pressure, but it can provide a useful way of measuring future weather conditions in a particular locality.

Jet Stream
The Northern Hemisphere polar jet stream is diverted southward by the Rocky Mountains, which may also create a trough (or area of lower pressure) to the east.

Winds from the lands of cold bring fruit of ice. Wind from the right hand of the west is the breath of the god of sand clouds.

This Native American proverb reflects a geographical reality of many tribes living in the western United States. Arctic air sweeps down into the Great Plains from Canada in the winter, and winds from the desert areas farther west bring dry, hot air.

When the wind is blowing in the North,
no fisherman should set forth.
When the wind is blowing in the East,
'tis not fit for man nor beast.
When the wind is blowing in the South,
it brings the food over the fish's mouth.
When the wind is blowing in the West,
that is when the fishing's best!

Cold northerly winds typically accompany a low-pressure system that brings stormy weather. As the low approaches, gusty easterly winds will usually pick up. In summer these winds are warm, dry, and dusty; in winter they can be bitterly cold. Northerly winds are typically unpleasant as well; they follow around a low-pressure area and tend to be cold and blustery. By contrast, southerly winds bring milder temperatures and favorable weather. The saying suggests that such conditions make it easier for larger fish to hunt for food; as a result, they are more active and easier to catch. But westerly winds are the most favorable weather for fishing because they tend to both bring fair weather and stay constant for a longer time.

Lows passing to the north of the observer are different from those passing to the south. In the former case, the wind typically blows from the east, then shifts to a southerly direction, possibly accompanied by light precipitation; then after the warm front passes, the winds shift again to the northwest or west as the cold front passes. A low passing to the south starts off much the same way with easterly winds, but will shift to a northerly direction in a short time. As the low passes due south, it will typically produce steady precipitation. As the low moves off to the east, though, skies clear and the winds will gradually shift to the west.

When rain comes before the wind,
dories, gear, and vessel mind.
When wind comes before the rain,
soon you'll make the set again.

Rain that comes before wind often means that a front is approaching that will bring unsettled weather for a day or two. On the other hand, wind that comes before rain may represent a downdraft from scattered showers that are moving in and will be more likely to blow over after a few hours.

USING AN ANEMOMETER

Anemometers are instruments used to measure wind velocity and pressure. Because wind velocity and pressure are related, information about one will also provide information about the other. The term comes from the Greek word *anemos*, meaning "wind."

The simplest type of anemometer is distinguished by three or four hemispherical cups mounted on each end of a pair of horizontal arms that lie at equal angles to each other. The cups are designed to catch the wind and cause the instrument to rotate aroud a vertical axis. By calculating the number of turns made at any given time, it is possible to determine the velocity of the wind.

CONSTRUCTING A WEATHER VANE

You'll need these materials:

- A long wooden dowel (about the length of a broomstick) or a broom handle
- An aluminum baking dish
- A 12-inch (30-cm)-long piece of wood or a sturdy ruler
- Nails
- Metal washer
- Hammer
- Glue
- Small saw
- Wire (for mounting)
- Scissors (strong enough to cut aluminum)

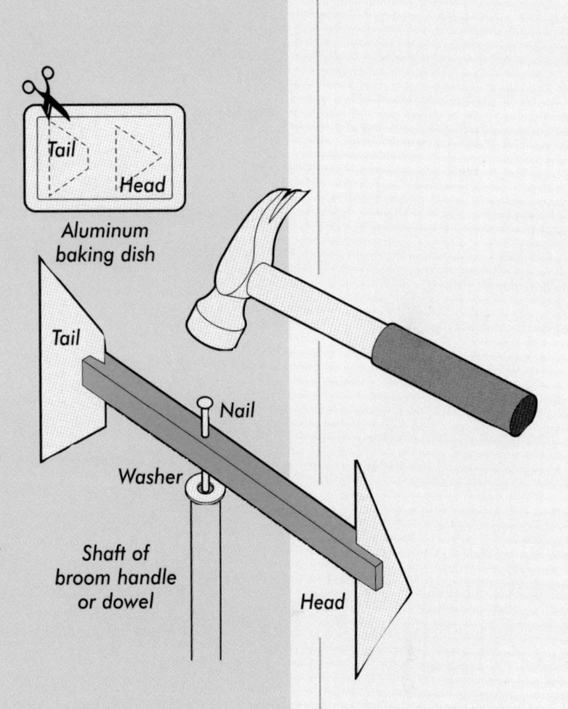

Cut the flat piece of wood with a saw to create a vertical slit about ½ inch (12.7 mm) deep at each end of the stick. Hammer one nail all the way through the stick at the midpoint of the top of the stick (exactly halfway across). Keep turning the wood around the nail several times until the stick turns easily. Now cut the head and tail out of the aluminum dish. Glue the head into the slot at one end of the wooden stick and the tail into the other end. You have created a weather vane.

Once the glue is dry, attach the weather vane to the long wooden dowel. Place the metal washer on the end of the dowel and then hammer the nail through the wooden stick and into the dowel. Make sure that the vane moves freely and easily around the nail. You can now mount the weather vane outside. You may want to use a fence and secure it by using wire. Try to get the vane as high above the fence as possible while making sure it is still well secured. The head of the pointer will always indicate the source of the wind. Do not make the mistake of thinking that the pointer tells you the direction in which the wind is blowing.

WHIRLWINDS

A whirlwind is defined as any rotating air mass generated in a storm system. They have come to mean mostly smaller spinning atmospheric disturbances and are also known as dust devils or dust whirls because they are produced in arid regions with a lot of sand in hot, dry weather. Solar radiation is mainly responsible for these phenomena.

The air mass just above ground becomes overheated; as the hot air rises, it assumes the form of a cylindrical column and sucks up the dust, debris, and sand on the ground. Whirlwinds can reach a height of 100 to 300 feet (30 to 91 m), but some dust devils have attained altitudes as high as 5,000 feet (1.5 km). Waterspouts are whirlwinds that occur over water.

Wind in the East, the fish bite the least. Wind in the West, the fish bite the best.

A steady westerly breeze and a light cloud cover on a mild day generally will prove congenial for the angler. The fisherman might be better off staying at home if the surface of the water is flat and calm or if there is a cold, gusting wind that brings squally showers from the east or north. Wind and clear skies alone, however, will not assure a good catch. Many other factors come into play, including slight changes in atmospheric pressure and water temperature.

If the wind is coming from the direction of the rainbow, the rain is heading toward you. Conversely, if the rainbow is in the opposite direction, it has passed you.

A rainbow is formed by the effect of sunlight on clouds, which separate the light into the colors of the spectrum. If the wind is blowing in your direction, the moisture-bearing clouds are heading your way, signaling that more rain is in store. If the wind is blowing in the direction of the rainbow, the rain clouds are dissipating and the storm has passed.

He who pays attention to the wind never sows his seeds; he who watches the clouds never harvests his crop.

This German saying seems to counsel caution about becoming preoccupied by the weather to the exclusion of the everyday tasks necessary to make a living. A farmer who waits for perfect conditions to prevail for planting or harvesting will probably come to grief.

A dead calm often precedes a violent gale.

While it is true that a calm comes before the storm, sometimes the storm comes without the calm. Thunderstorms are generated out of rising hot air carrying water vapor that at higher altitudes condenses and forms moisture-laden clouds. The rising air packets form a partial vacuum that sucks cold air from above it. These cooler downdrafts are responsible for causing the rain to fall (or hail or sleet or snow), while the turbulence so characteristic of these storms comes about because

WIND GODS

Throughout history many cultures have worshipped wind gods, but the Greeks had an elaborate classification system. The most prominent was Aeolus, who was capable of producing winds of all kinds, depending on the instrument he used or his mood. When he played his harp, a breeze would ruffle the leaves; but if he chose to blow through his conch shell, a catastrophic gale would result. Ordinarily the wind was safely locked away in a whistling cavern. But there was a division of labor: Boreas was put in charge of the north wind (from him we derive the word *borealis*); Zephyrus presided over the mild west wind (whence the word *zephyr*); and Eurus (whose surname Argestus meant "the bearer of brightness") ruled over the southeast wind. These were the original wind gods cited by the poet Hesiod (c. 700 B.C.), but over time their number expanded to include gods of the south wind, northeast wind, and more. Boreas ruled over the coldest and most powerful wind. Zephyrus was known as protector of plants (because his breezes brought humidity). Zephyrus was depicted as a handsome young man; Eurus had a melancholic face; Notus, who ruled the south wind, was a feared figure, especially by seafarers, because of the storms he could bring.

Chinese Wind God

The Chinese worshipped Fei Lian as the god of the wind, a responsibility he shared with another wind god, Feng Po. Many ancient cultures had wind gods. In Mycenaean Knossos, for instance, the divine family of winds had no less than eight members who inspired a cult of winds.

of the combined action of the cold downdrafts and warm updrafts. The partial vacuum also drags in air from all sides of the storm front. The calm that arises in this way, however, is more typical of a single-cell thunderstorm. When several thunderstorms move in together—each storm is known as a cell—there is so much air in motion that there is no calm before the storm hits.

Whirlwinds are caused by the devil.

The belief that whirlwinds are the work of the devil, a belief once held in Iran, refers to the effect of the spring winds in that country. They arise in unstable conditions between the end of winter and the beginning of summer, from centers of low pressure that develop on the Iranian plateau and in western Central Asia. These systems generate heavy downpours, dust storms, and sandstorms. This season is known for the frequency of whirlwinds that are filled with dust. These winds are produced in the afternoon because of sharp differences in temperature.

The north wind is best for sowing seed, the south for grafting.

This English saying is not a very reliable guide for gardeners, although it is true that sunlight and wind can help germination. Gardening experts recommend spring and early summer as the best time for germinating seeds. In climates where seedlings may need protection from the cold and wind, it is best to cover them with leaves or straw.

Every wind has its weather.

Different winds do in fact presage different weather conditions. Wind is an indication of two major weather developments: temperature and moisture advection. In meteorology, advection refers to the horizontal transport of an atmospheric property (such as heat or moisture) by the wind. If temperatures are warmer upwind, then it would indicate that the wind will bring warmer temperatures downwind. Similarly, if the temperatures are colder where the wind is originating, you could expect that temperatures downwind will drop. The same holds true for humidity.

Some winds are more important than other winds—at least as far as meteorologists are concerned. Synoptic winds are the most important because these are winds that are associated with warm and cold fronts, which make up most day-to-day weather. (The word *synoptic* is derived from two Greek words meaning "together with" and "visible," hence, meaning "seen together.")

A western wind carrieth water in his hand; when the east wind toucheth it, it shall wither.

The validity of this saying, like so many other weather proverbs, depends on where you are. If for example you live in the western United States, the saying makes sense because moisture-laden clouds do blow in from the Pacific and from the southwest. Easterly winds blowing over desert regions, on the other hand, will bring hot, dry conditions. For people living on the East Coast of the United States, however, the saying offers little guidance. In their case the rain moves in with winds from the Atlantic Ocean (that is, from the east) or from the south (the Gulf of Mexico). Anyone who lives in or near the Cascade Mountains and the Sierras in the western part of

FAMOUS WINDS

Some winds are classified based on their particular geographical, topographical, or seasonal characteristics.

- Valley winds are created by the warming of mountain slopes during the day.
- A Chinook, also known as bitter wind, is a warm, dry wind that descends on the leeward side of a mountain.
- Santa Ana winds are hot, dry winds that occur over plateau regions in the western deserts of the United States.
- A foehn is a warm, dry wind that blows down a side of a mountain. The most common one is in the Alps.
- The scirocco is a hot, dry wind that blows in southern Europe and North Africa.
- In the Middle East, hot, dry winds known as the Sharav and as the Khamsin blow from the Arabian desert.

BEAUFORT WIND SCALE

Force	Miles per hour (kph)	Description
0	0–1 (0–1)	Calm
1	1–3 (1–5)	Light air
2	4–7 (6–11)	Light breeze
3	8–12 (12–19)	Gentle breeze
4	13–18 (20–29)	Moderate breeze
5	19–24 (30–39)	Fresh breeze
6	25–31 (40–49)	Strong breeze
7	32–38 (50–61)	Moderate gale
8	39–46 (62–74)	Gale
9	47–54 (75–88)	Severe gale
10	55–63 (89–102)	Storm
11	64–72 (103–117)	Violent storm
12	73–82 (118–134)	Hurricane

LAND AND SEA BREEZES

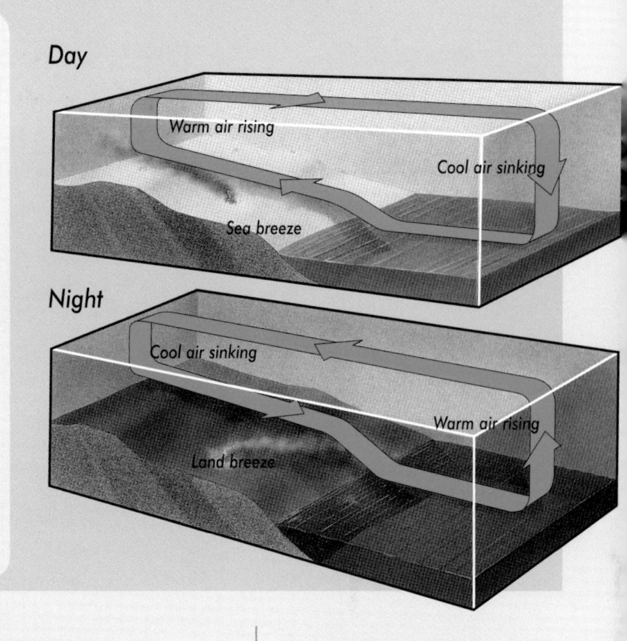

the country would also have cause to ponder why so much praise is lavished on westerly winds, since those are the very winds (along with southwesterly winds) that bring in clouds that pummel the mountains with snow and rain.

The south wind warms the aged.

In many cases, this saying is true enough: Retirees often seek the warmth of southerly climates. However, in summer and autumn, southerly winds,—far from being harbingers of benign weather—can produce very wet and stormy weather, especially hurricanes, from the south Atlantic and the Gulf of Mexico.

*Trace in the sky the painter's brush,
the winds around you soon will rush.*

The painter's brush of this saying refers to cirrus clouds. These high-level clouds are composed of ice crystals, and their appearance often precedes the approach of a storm system accompanied by heavy precipitation and blustery winds.

Land and Sea Breezes

Sea breezes originate from the sea and blow towards land. On warm, sunny days the land heats more quickly than the nearby waters; warm air off the land rises and draws in the cooler air from the sea to replace it. At night, however, the pattern reverses because the sea maintains its warmth longer than land, drawing the cooler air towards the sea—a phenomenon called a land breeze.

WIND FACTS

- Hurricane strength is calculated in wind speed. The weaker storms are classified as Category 1, with speeds of 74 to 95 m.p.h. (119 to 153 kph), and the most powerful as Category 5, with winds exceeding 155 m.p.h. (249 kph). Even Category 1 hurricanes can be devastating. Hurricane Agnes, a Category 1 storm that struck the United States in 1971, killed 122 people in Florida and the Northeast.
- The strongest wind on record was 318 m.p.h. (512 kph) and was produced by a tornado that tore through parts of Oklahoma on May 3, 1999.
- Super Typhoon Nancy, which struck Japan in 1961, reached an estimated sustained wind speed of 213 m.p.h. (343 kph).

Winds that change against the sun,
are always sure to backward run.

This saying refers to a backing wind that shifts in a counterclockwise direction—against the sun—and is associated with approaching low-pressure systems that will bring rain and gusty winds.

Cyclones generally move as a whole to the westward, curving in to the northward, in northern latitudes; and to the westward, curving to the south, in southern latitudes.

Cyclones, which can be tropical hurricanes or extratropical (in that they form over land or water outside of the tropical zone), or even polar, are defined as a region of low atmospheric pressure with inward-spiraling winds that spin counterclockwise in the Northern Hemisphere and clockwise in the Southern Hemisphere. So this saying is accurate only in the former case. Hurricanes, for example, usually develop in the warm waters of the Caribbean or south Atlantic, then move west in the direction of the Gulf Coast of the United States, Central America, and Mexico, and/or north up the Eastern Seaboard of the United States.

A frequent change in the wind,
with agitation in the clouds, denotes a storm.

Shifting winds generally signal a dramatic change in the weather, but the type of change will usually depend on atmospheric pressure, temperature, and other factors. In the United States, for example, winds from the east with a falling barometric pressure indicate foul weather, while winds shifting to the west indicate clearing and fair weather; but there are exceptions. A southerly or southeasterly wind, together with a falling barometer, means that a storm is approaching from the west or northwest and that its center will pass near or north of the observer within 12 to 24 hours, with winds shifting to the northwest. A wind from the east or northeast with a falling barometric pressure means that a storm is coming in from the south or southwest and that its center will pass near or to the south of the observer within 12 to 24 hours, with winds shifting to northwest. The faster the pressure is dropping, the more rapidly the storm will approach and the more intense it will be.

HURRICANE FORMATION

Hurricanes are Earth's most powerful cyclones. They can have a long life cycle: Some have survived for up to two or three weeks, diminishing in strength to a tropical storm and strengthening again along the way. That cycle often begins in the waters west of Africa that have acted as a nursery for tropical storms over the last century. Hurricanes quickly lose strength and die if they cross cooler water, which deprives them of the heat that will maintain their power.

During hurricane formation, hot air over tropical oceans begins to rise, producing tall, mainly cumulonimbus clouds. With the depletion of air at the ocean surface, a low-pressure system forms, which sucks in air from the surrounding area. That air is warmed in turn, rises, and produces more tall clouds. The clouds begin to join in spiral bands, generating thunderstorms. Higher atmospheric winds act to keep the storm organized.

Hurricanes are cyclonic systems made up of a large, rotating system of clouds, winds, and thunderstorms. The life cycle of a hurricane begins as a tropical cyclone forms around areas of low pressure, powered by heat produced as moist air rises and condenses. A tropical depression forms when a group of thunderstorms coalesces under the right atmospheric conditions for a long enough time to organize into a single system with winds near the center between 20 and 34 knots (23 to 39 m.p.h., or 37 to 63 kph). They form in regions of high humidity that reduce the amount of evaporation in clouds and maximize the heat released, without which these storms cannot develop. The ocean water itself must be warmer than 80°F (26.5°C).

When a tropical depression intensifies so that its maximum sustained winds reach 35 to 64 knots or 40.2 to 74 m.p.h. (65 to 119 kph), it becomes a tropical storm and is honored with a name. If the storm assumes a circular shape, it can turn into a more powerful hurricane. To be defined as a hurricane, a storm must have winds that exceed 64 knots and circulate about their centers. They are also distinguished by a dark spot in their center, called an eye.

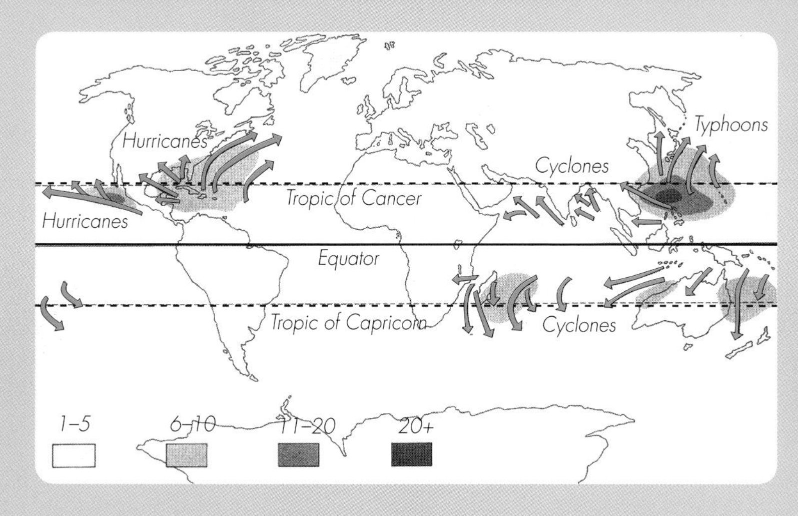

Left: *This map shows the typical tracks and frequency of tropical cyclones over a 20-year period.*

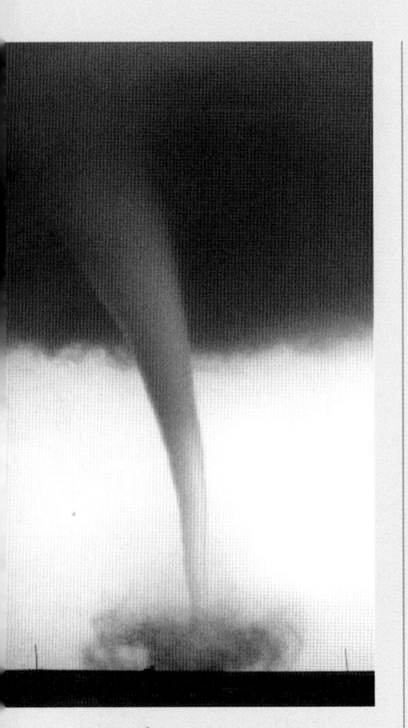

Tornadoes

Tornadoes are the most violent storms on Earth, but most of them do not cause destruction or loss of life. Ninety percent of tornadoes last 20 minutes or less. Just two percent of these storms, which can endure up to an hour, account for 70 percent of the deaths.

TORNADOES

These funnel-like clouds are among the most destructive storms, though only very few cause real damage. Usually the product of violent thunderstorms, they can achieve wind speeds unequaled by any other type of storm. Although tornadoes are a frequent occurrence, scientists still have not quite understood the way they form or how they achieve their devastating power. Tornadoes are probably the subject of more myths, misconceptions, and folklore about how to escape them than any other type of storm.

A town will be protected from tornadoes if it is protected by a river.

There is no reason to think that any location is protected from tornadoes, regardless of its perceived geographical advantage. Tornadoes have proven quite capable of crossing rivers and even canyons. In 1840 a tornado crossed the Mississippi River, briefly turning into a waterspout, before resuming its course on land, killing hundreds in its path. The Great St. Louis Cyclone of 1896 simply jumped the Mississippi River. The Tri-State Tornado of 1925 did the same thing, leaving almost 700 dead. Tornadoes have crossed rivers of all kinds and widths. If there is one place that is probably safe from a tornado, it is most likely Antarctica.

Tornadoes have picked people and items up, carried them some distance, and then set them down without injury or damage.

While it seems improbable that people, animals, or breakable objects could survive a tumultuous airborne journey, the fact is that they do. Many instances have been recorded where people and animals have been transported up to a quarter mile (0.4 km) or more without serious injury. Even fragile items, such as sets of fine china or glassware, have been blown from houses and then recovered, miles away, completely intact. Papers carried by tornadoes have been found as much as 100 miles (161 km) away. It is just that such occurrences are more likely to be the exception and not the rule. The vertical winds in tornadoes can be strong enough

TORNADO FORMATION

Tornadoes develop in severe thunderstorms known as supercells, which persist longer than single-cell thunderstorms because they are sustained by more "fuel" (moisture, wind, and cooler downdrafts and warmer updrafts). Thunderstorms require three ingredients to develop: unstable air, a mix of warmer and cooler air, and a lot of moisture in the atmosphere. The same factors that keep a supercell going are also responsible for producing more tornadoes. (Winds coming into a storm begin to swirl and form the distinctive funnel that typifies most tornadoes. As the air in the funnel spins faster and faster, it creates a localized low-pressure area that sucks in air and debris.

The severe thunderstorms that unleash the strongest tornadoes tend to occur in areas where cold, dry, polar air masses meet warm, moist tropical air. Those conditions are especially prevalent in the U.S. Midwest (which includes the Mississippi, Ohio, and lower Missouri river valleys) and the southeastern United States, where humid air moves in from the Gulf of Mexico and collides with colder air from Canada in spring. This region has been dubbed Tornado Alley. The warmer air, usually above 75°F (24°C), and the cooler, faster-moving air is given added energy by a fast-moving jet stream above.

The jet stream, which flows above 30,000 feet (9 km), helps develop tornadoes by creating a vacuum below it, making it easier for the rising air to continue to rise at higher altitudes.

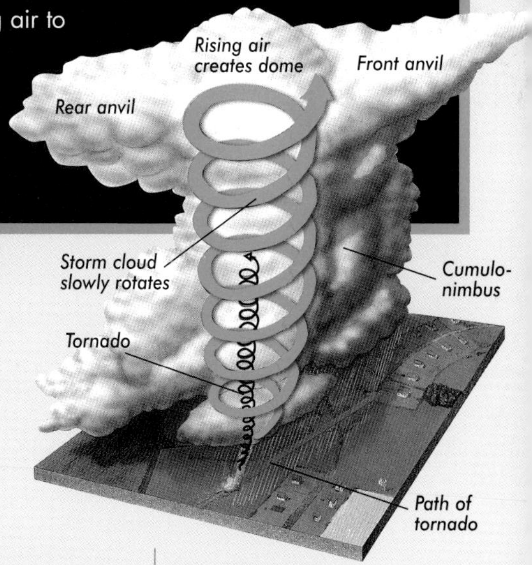

to lift even heavy objects if they have a large face to the wind, or flat sides such as roofs, walls, trees, and cars. If the object is relatively light, the winds can temporarily lift them tens of thousands of feet (several kilometers) high.

Hiding under a freeway overpass will offer protection from a tornado.

A driver would be ill-advised to stop under an overpass in the event of a tornado. There are a number of cases where people have been killed by flying debris under such structures. An overpass is especially dangerous because it acts as a wind tunnel and may actually serve to attract more debris. Moreover, the winds in a tornado tend to gain speed with height; under such conditions, winds can easily exceed 300 m.p.h. (483 kph). It is much more prudent to abandon the vehicle and take refuge in a ditch or other low-lying location.

Rising air creates dome
Front anvil
Rear anvil
Storm cloud slowly rotates
Cumulo-nimbus
Tornado
Path of tornado

Tornado Formation
Sometimes tornadoes can resemble a cloud moving along the ground, rather than a funnel. Rain can often wrap around the cloud, obscuring its characteristic form.

The strongest tornadoes are the biggest and fattest ones.

Although there is some statistical evidence to suggest that wider tornadoes may cause more damage, that does not mean they are necessarily stronger as measured on the Fujita Tornado Intensity Scale. This scale differs from scales measuring the intensity of other types of storms. For instance, hurricanes are classified by the strength of their winds based on the Saffir-Simpson Scale (from Category 1 up to the strongest, Category 5). Tornadoes, on the other hand, are assessed by the damage they do after the fact. This is because it is impossible to determine the "true" wind speeds at ground level in most tornadoes, and because winds of different strength can do similar damage, with the extent of the damage varying from block to block and building to building.

The scale, sometimes called the F Scale, was named for its inventor, Dr. T. Theodore Fujita (1920–1998), and has been used widely around the world for the last three decades, with certain modifications. The scale actually goes up to F12, but that would require wind speeds exceeding Mach 1.0, or around 761 m.p.h.

TORNADO SAFETY TIPS

- In tornado-prone regions, it is important to stay tuned for tornado warnings. Purchase a tornado/severe weather kit that contains a battery-powered radio, a flashlight, a first aid kit, food, and a blanket. Set up an evacuation plan. You should also know of the location of the nearest shelter.
- A tornado watch is issued when conditions are favorable for thunderstorms that can generate tornadoes. A tornado warning is issued when a tornado has been identified and it poses a danger to your locality. At this point you should find shelter immediately.
- No room in a mobile home is safe. All mobile homes should be evacuated in the event of a tornado warning. If you live in a sturdy house, you should take shelter in the basement, preferably in an interior room away from walls exposed to the outside. In a house without a basement, you should find shelter in an interior room. Once inside the shelter, you should kneel down with your head against a wall and cover your head with your arms.
- If a tornado is near and you are caught outside without enough time to find shelter quickly, you should take refuge in a low place, such as a ditch, and lie down. Continue to remain as low as possible to minimize exposure of your body to flying debris. Never seek shelter under a bridge or a highway. Do not seek shelter or remain in a vehicle—these can be death traps; you are better off abandoning a vehicle. Avoid trees, just as you would in a thunderstorm.

(1,225 kph), a speed far in excess of any ever documented on Earth. The highest wind speed attained by a tornado was about 318 miles per hour (512 kph)—registered as an F5. So while a "fat" tornado may not be stronger than one with a thinner funnel cloud, it might wreak more damage in its path simply because its path is more extensive. Tornadoes may assume a variety of shapes—a wedge tornado appears to be as wide as it is tall (measured from ground to ambient cloud base), and a rope tornado is snakelike in form. (*Wedge* and *rope* are slang terms.) But their shape has nothing to do with their strength.

A tornado can be outrun, especially in a vehicle.

Trying to outrun a tornado, even behind the wheel of a Ferrari, would be a mistake. For one thing, tornadoes can move at up to 70 m.p.h. (113 kph) or more—there are recorded speeds of over 300 m.p.h. (483 kph)—so they could easily overtake most cars, even those driving at high speeds. For another, they can shift directions erratically and without warning, so it would be impossible to know how to drive toward safety. Drivers are generally advised to leave their cars and find shelter on foot. Some veteran tornado observers, however, argue that under certain circumstances it might make sense to attempt to flee a tornado in a vehicle. The driver is cautioned to watch the storm closely for a few seconds relative to a fixed object in the foreground, such as a tree. If the tornado looks as though it is moving to the right or left, then it is a safe assumption that it is not a direct threat. The driver should try to escape right if the tornado is moving to the left, and left if it is moving to the right. If the tornado appears to stay in place or is growing larger, however, moving neither to the right nor to the left, it is heading for the driver, in which case abandoning the vehicle—quickly—is the only safe alternative.

Tornadoes are more likely to hit a mobile home park.

It is a fallacy to believe that tornadoes would single out mobile home parks more than any other residential (or commercial) location—tornadoes are too indiscriminate—but mobile homes are certainly more vulnerable. A mobile home will suffer far more damage if it is in the path of a tornado than a frame house built on a foundation.

TORNADO FACTS

- In a typical year, more than 900 tornadoes occur in the United States, the highest frequency reported for any nation.
- The deadliest tornado—an F5—ever recorded is the "Tri-State" Tornado of March 18, 1925, which killed an estimated 695 people, as it followed a course 219 miles (352 km) long across parts of Missouri, Illinois, and Indiana. This tornado claimed 234 lives just in Murphysboro, Illinois—the most ever for one locality.
- The biggest outbreak of tornadoes was the "Super Outbreak of 1974," when a total of 147 tornadoes touched down in 13 American states and in Canada, killing 310 people in the United States, and 8 in Canada.

Tornadoes don't happen in the mountains.

Tornadoes do occur in the mountains. Damage from a tornado has been reported above 10,000 feet (3 km). Tornadoes have barreled across mountain chains including the Appalachians, the Rockies, and the Sierra Nevada. In August 1911 the Salt Lake City tornado climbed up one side of a canyon and climbed up half the distance of the other side before finally dissipating. In 1987 an especially violent tornado crossed the Continental Divide in Yellowstone National Park.

Rain always comes before a tornado.

In spite of the fact that most tornadoes are spawned from thunderstorms, rain does not necessarily precede a tornado. Tornadoes are so unpredictable that many different weather conditions may presage them, depending on the storm and the circumstances under which the storm is created. It will also depend on the direction the storm is moving. Unusually large hailstones may indicate the presence of a major thunderstorm and a tornado, but it is not a reliable predictor.

Tornadoes can skip.

Tornadoes cannot skip because, by definition, the funnel cloud of the tornado must remain in constant touch with the ground. There is some dispute as to whether multiple touchdowns by the same funnel cloud constitute one or many tornadoes.

PREDICTING WEATHER USING THE INTERNET

AccuWeather
www.accuweather.com

Environment Canada
www.weatheroffice.ec.gc.ca

National Weather Service
www.nws.noaa.gov

Stormtrack
www.stormtrack.org

World Meteorological Organization
www.worldweather.org

Weather Underground
www.wunderground.com

The Weather Channel
www.weather.com

National Hurricane Center
www.nhc.noaa.gov

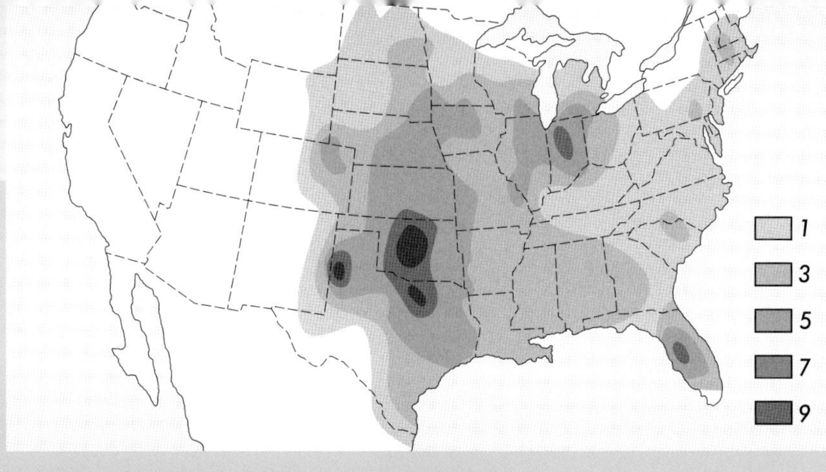

Average number of tornadoes per 10,000 sq. mi. (25,000 sq km) per year in the United States.

WHERE DO MOST TORNADOES OCCUR?

In absolute numbers, the United States has the most tornadoes by far. It also has the most-violent tornadoes (about 10 to 20 per year). No region of the United States is more susceptible than the area known as Tornado Alley. But all regions of the United States are at risk for tornadoes, and they occur frequently in the Plains area between the Rocky Mountains and the Appalachians, including the states of Oklahoma, Kansas, Arkansas, Iowa, and Missouri, as well as northeastern Texas, eastern Colorado, northern Louisiana, central and southern Minnesota, southwestern Indiana, and parts of Nebraska. Tornado Alley even encompasses small portions of Tennessee, Kentucky, and Wisconsin.

Canada reports the second largest number of tornadoes (about 80 to 100 annually). The United Kingdom has the most tornadoes per land size—about 33 a year, most of them weak. Russia may have many tornadoes, but the data is not available. About 20 tornadoes are reported in Australia each year, mostly in New South Wales, although the actual number is probably higher. South Africa and Argentina also experience tornadoes—about 200 each over the 40-year period from 1930 to 1979.

Tornadoes may occur in the middle of the night and even during the winter.

Tornadoes are most commonly associated with thunderstorm activity, which occurs primarily in warmer weather. They can in fact occur even at night or in colder months, although the likelihood of their occurrence diminishes. The most likely time for a tornado is on summer afternoon, when thunderstorms tend to develop.

Tornadoes sound like trains.

People have compared their sound to a train because they can produce a continuous rumble. Tornadoes can also make a sound that recalls a waterfall, and once they begin barreling through an urban area, they can produce a terrifying roar.

Expect tornado weather when dogs eat grass.

Whether the behavior of any animal can predict meteorological phenomena is a subject of debate. However, observation over centuries offers some tantalizing evidence that animals do react in advance of storms and even earthquakes. Predictive behavior has been observed more often in wild animals than in domesticated ones. One reason that chickens and birds in particular may react to approaching weather is because they are very sensitive to changes in atmospheric pressure, in part because of their hollow bones.

The distinctive behavior of animals is usually seen about 24 hours prior to a storm or other natural disaster. Cats will hide out, and dogs tend to get nervous, clingy, or aggressive. Wild animals may flee their natural habitat. Many dog owners maintain that their pets have sensed earthquakes in advance and forced them to leave their homes when they felt one was about to occur. In 2006 authorities in Clark County, Oregon, reported that dogs at a pound were "giving off a different vibe" than usual just hours before a 3.8-magnitude earthquake occurred not far away. "It was one of those days where everyone was wanting to bark," a worker at the facility stated. "They were barking their heads off. This was far more anxious barking. I don't know if they knew what was happening. It was unusual for them." There is, however, no scientific evidence to confirm that by eating grass, dogs are capable of predicting the approach of a thunderstorm or tornado—at least not yet.

WEIRD PRECIPITATION

Tornadoes and other intense windstorms have deposited some of the most unexpected and sometimes alarming objects and creatures miles from where they were suddenly scooped up for their unanticipated odyssey through the heavens. There have been cases of people and cows being dropped down 10 miles (16 km) away.

Parts of England have witnessed rainfalls of snails, and the United States has been struck by falls of eels. There have been many documented cases of frogs, spiders, and snakes dropping down from the sky. Several "fish storms" have been reported in Australia. In India, fish weighing up to 8 pounds each (3.6 kg) came down from the sky.

To avoid the impact of a tornado, you should open the windows to equalize the pressure inside and outside the house.

This is a dangerous fallacy. Opening the windows will only increase the danger to the inhabitants of being struck by flying debris. In any case, if a tornado should strike a house, it will blast the windows open anyway.

A tornado can drive a straw through a telephone pole.

Tornadoes do, in fact, have the capacity to turn even everyday objects into missiles. Although some of the reports of damage from bizarre objects borne by tornadoes is folklore, storm winds have been known to strip asphalt

WATERSPOUTS

There are two types of waterspouts: weaker "fair-weather waterspouts" and much stronger "tornadic waterspouts." Like tornadoes, waterspouts have a similar basic structure typified by upward-moving air. The funnel cloud owes its origin to the action of rushing winds at the water surface; as the winds intensify, they swirl into a vortex and then begin to rise up. Though they begin at the surface, the funnel clouds don't actually seem to be touching the water. When the wind speeds reach around 40 m.p.h. (65 kph), the wind causes spray to form a circular pattern—the spray vortex. The funnel cloud that emerges from this process extends all the way from the ocean to the parent cumulus cloud. The vortex can only be seen at an altitude high enough for the lower pressure to condense the water vapor into water droplets. That is why it looks as if the sprouts aren't touching the water.

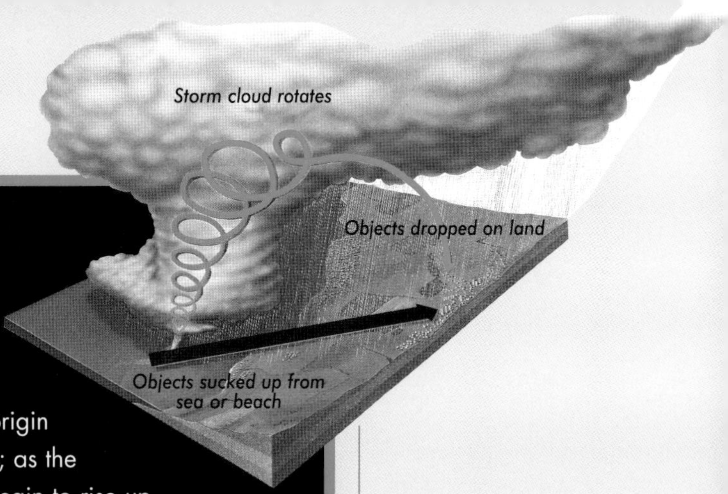

Waterspouts

Waterspouts are essentially weak tornadoes over water that usually take the form of funnel-shaped clouds. They tend to appear in association with a cumuliform cloud. Waterspouts can attain speeds up to 190 m.p.h. (306 kph).

pavement from roadbeds and fling wooden splinters with such force that they can pierce brick walls. There have been cases where the winds have bent trees, creating cracks which they fill with debris before bending them back, locking the trees in place. In spite of many studies undertaken on tornadoes, scientists are still uncertain about what actually occurs inside their vortex.

Hurricanes and tropical storms can produce tornadoes.

Hurricanes that make land may or may not cause tornadoes. Hurricane Andrew, one of the most devastating in U.S. history, struck in 1992, spawning several tornadoes in several Southern states; but once it had crossed the Gulf of Mexico, it produced only one tornado in Florida. Tropical storms and tropical depressions are also capable of causing tornadoes, though to a lesser extent. When these tornadoes do occur, they usually form in small supercells. The strength of the parent storm does not invariably predict the power of the tornado. Relatively weak hurricanes like Danny (1985) have caused significant supercell tornadoes well inland. The strength of the tornadoes generally depends on the extent of the wind fields within a tropical cyclone.

Resources

Books

Burt, Christopher, and Stroud, Mark. *Extreme Weather: A Guide and Record Book*. New York: W.W. Norton & Company, 2004.

Cerveny, Randy. *Freaks of the Storm: The World's Strangest True Weather Stories*. New York: Thunder's Mouth Press, 2006.

Cox, John D. *Weather for Dummies*. Hoboken, NJ: For Dummies, 2000.

Dennis, Jerry. *It's Raining Frogs and Fishes: Four Seasons of Natural Phenomena and Oddities of the Sky (Outdoor Essays & Reflections)*. New York: Harper Paperbacks, 1993.

Dunlop, Storm. *The Weather Identification Handbook: The Ultimate Guide for Weather Watchers*. Guilford, CT: The Lyons Press, 2003.

Flannery, Tim. *The Weather Makers: How Man Is Changing the Climate and What It Means for Life on Earth*. New York: Atlantic Monthly Press, 2006.

Garriott, Edward B. *Weather Folk-Lore and Local Weather Signs*. Honolulu: University Press of the Pacific, 2001.

Geer, I.W. (ed.). *Glossary of Weather and Climate*. Boston: American Meteorological Society, 1991.

Gibbons, Gail. *Weather Forecasting*. New York: Aladdin, 1993.

Goldstein, Mel. *The Complete Idiot's Guide to Weather*. Indianapolis, IN: Alpha, 2002.

Flannery, Tim. *The Weather Makers: How Man Is Changing the Climate and What It Means for Life on Earth*. New York: Atlantic Monthly Press, 2006.

Hodgson, Michael. *Basic Essentials: Weather Forecasting*. Guilford, CT: Globe Pequot, 1999.

Lee, Albert. *Weather Wisdom: Facts and Folklore of Weather Forecasting*. New York: Congdon & Weed; Reprint edition, 1990).

LeMone, Margaret. *The Stories Clouds Tell*. Boston: American Meteorological Society, 1993.

Linden, Eugene. *The Winds of Change: Climate, Weather, and the Destruction of Civilizations*. New York: Simon & Schuster, 2007.

Lockhardt, Gary. *The Weather Companion: An Album of Meteorological History, Science, and Folklore*. New York: John Wiley, 1988.

Inwards, Richard. *Weather Lore*. London: Senate, 1994.

Passante, Christopher K., Bologna, Julia. *The Complete Idiot's Guide to Extreme Weather*. New York: Apha Books, 2006.

Rubin, Louis, Duncan, Jim, and Herbert, Hiram. *The Weather Wizard's Cloud Book: A Unique Way to Predict the Weather Accurately and Easily by Reading the Clouds*. New York: Algonquin Books, 1989.

Sloane, Eric. *Almanac and Weather Forecaster*. New York: Duall, Sloan, and Pearce, 1957.

Sloane, Eric. *Folklore of American Weather*. New York: Duall, Sloan, and Pearce, 1963.

Sorbjan, Z. *Hands-On Meteorology*. Boston: American Meteorological Society, 1997.

Streluk, Angella, and Rodgers, Alan. *Measuring the Weather: Forecasting the Weather*. Heinemann Educational Books, 2002.

Websites

Weather folklore and myths:

Hudson Valley (NY) Weather Folklore
www.xomba.com/hudson_valley_weather_forecasting_folklore

The Most Popular Myths in Science
www.livescience.com/bestimg/index.php?url=myths_lightning_strike_03.jpg&cat=myths

Nature's Calendar (Phenology page)
www.naturescalendar.org.uk/

The Old Farmer's Almanac
www.almanac.com/

Old Weather Predictions, Sayings, & Folklore
mysite.verizon.net/dfairlie/folklore.html

Shakespeare's Folklore and the English Holiday Cycle
www.endicott-studio.com/jMA03Summer/shakespeare.html

Storm Fax: Weather Folklore and Folklore Forecasts
www.stormfax.com/wxfolk.htm

Tornado and Storm Folklore
www.cyberlodg.com/mattdennis/ktc/folk.htm

Unusual Animal Behavior (Animals and Weather: news from around the world)
http://home.att.net/~thehessians/newanimal.html

Weather Lore (entry): Wikipedia
http://en.wikipedia.org/wiki/Weather_lore

Weather Folkore collection
http://members.aol.com/Accustiver/wxworld_folk.html

Weather World Folklore
www.islanderz.com/weather/folklore.html

National and international weather services:

Environment Canada
www.weatheroffice.gc.ca/canada_e.html

The U.S. National Weather Service
www.nws.noaa.gov

USA Today (Weather Page)
www.usatoday.com/weather/wresources.htm

Weather-tracking and general meteorology:

The American Meteorological Society
www.ametsoc.org/amsedu/dstreme/

Royal Meteorological Society (U.K.)
www.rmets.org

Science Daily
www.sciencedaily.com/news/earth_climate/

Stuff in the Air: The Science of Meteorology Online
www.stuffintheair.com/

Weather Cameras Across the United States and Canada
http://cirrus.sprl.umich.edu/wxnet/wxcam.html

Weather Resources
www.refdesk.com/weath1.html

Index

Page numbers in italic refer to illustration captions.

M

N

O

P

R

S

T

U, V

W

Y

Z

ACKNOWLEDGMENTS

Artwork:
Pages 13 Peter Sarson/Richard Chasemore; 14 Mark Franklin; 58 Mark Franklin; 59, 99, 106 Robert Brandt, 93 Gary Hincks; 133 Gary Hincks; 151 Peter Sarson/Richard Chasemore

All other illustrations by Simon Gurr.

Photographs:
Pages 8–9 Naturfoto Honal/Corbis; 10 Swim Ink 2, LLC/ Corbis; 16 Archivo Iconografico, S.A./Corbis; 20 Archivo Iconografico, S.A./Corbis; 24 Bettmann/Corbis; 38 Uli Wiesmeier/zefa/Corbis; 42 Sygma/Corbis; 43 Bettmann/ Corbis; 48 Reuters/Corbis; 53 Larry W. Smith/epa/Corbis; 67 Frank Earle/Tony Stone Images; 94–95 Robert Dowling/Corbis; 96 Hulton-Deutsch Collection/Corbis; 97 John Heseltine/ Corbis; 103 Jane Sweeney/Robert Harding World Imagery/ Corbis; 111 Bettmann/Corbis; 114 Chinch Gryniewicz; Ecoscene/Corbis; 121 George D. Lepp/Corbis; 126 Philip Wallick/Corbis; 127 Andrea Merola/epa/Corbis; 129 Historical Picture Archive/Corbis; 134–135 Bettmann/ Corbis; 141 Bettmann/Corbis; 143 Corbis; 144 DiMaggio/ Kalish/Corbis; 147 Werner Forman/Corbis; 149 Sheila Terry / Science Photo Library; 150 Jim Zuckerman/Corbis; 157 Burstein Collection/Corbis; 162 Eric Nguyen/Corbis